ECONOMICS AND THE THEORY OF GAMES

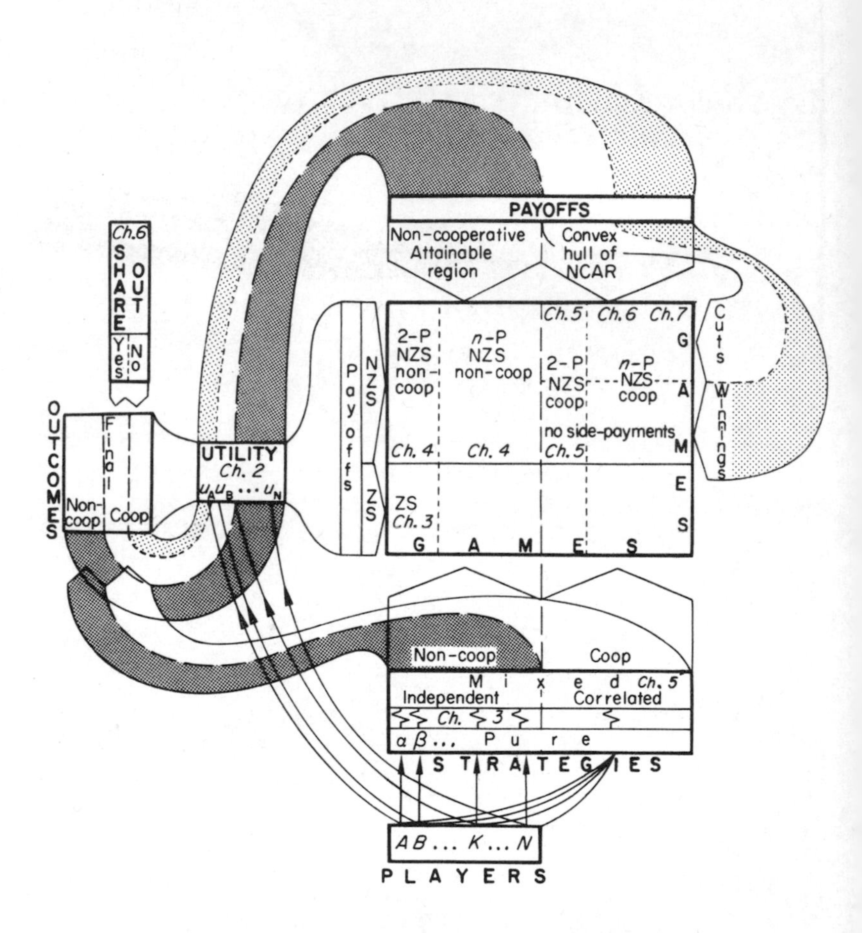

The structure of game theory

Economics and the Theory of Games

Michael Bacharach
Student of Christ Church, Oxford

M

First published 1976 by
THE MACMILLAN PRESS LTD
London and Basingstoke
Associated companies in New York Dublin
Melbourne Johannesburg and Madras

ISBN 0 333 19592 2 (hard cover)
333 19593 0 (paper cover)

Typeset, printed and bound
in Great Britain by
Redwood Burn Limited

Contents

Preface

This book has two aims: to expound game theory for students who are not mathematical specialists, and to apply this theory in economics.

In spite of the second of these aims, the book is intended not only for economists but also for other social scientists, for management trainees, and indeed for anyone who wants to find out about game theory. Non-economists will have little difficulty in following the economic parts of the book; in any case, sections dealing with economic applications can be skipped with hardly any loss of continuity – although it would be a pity to do so since much of the richness and colour of game theory resides in these applications.

Game theory includes not only the solution of games themselves, but also the von Neumann and Morgenstern theory of cardinal utility on which the analysis of games rests. The range of application of this utility theory is vast, and stretches far beyond games, for it is nothing less than the theory of rational individual choice when *risk* is present – that is, when only probabilities are known for the consequences of choices.

Games may be divided into: two-person games – zero-sum and nonzero-sum, cooperative and non-cooperative; and *n*-person games. Two-person zero-sum games have received far more treatment in the literature than other games. They have limited direct application, but are of great intrinsic interest as a paradigm case of multi-person decision-making. The theory of these games is also an essential analytical step in developing the rest of game theory. The theory of other types of games has many, vitally important applications, in economics, politics, strategic studies and elsewhere; some of those dealt with in detail in this book are oligopoly, markets with externalities, wage bargaining and competitive equilibrium.

Game theory directly confronts essential features of these problems which traditional analysis tries to ignore – namely the *interdependence* of people's decisions and the *uncertainty* under which they have to make them. The fast-increasing interest in game theory is in part due to an appreciation of the way in which, in approaching these problems, it takes the bull by the horns; so far from banishing interdependence and uncertainty, its very starting point is to ask what rational decisions *are* in their presence.

This book is less than an advanced treatise on game theory, but more than an introduction. It is fairly comprehensive, and it is quite rigorous while being undemanding mathematically. The reader is led, if not by the hand, at least by the elbow. The book has nevertheless to be read with care, both because some things are said only once, and because at times the subject is just not easy. But as much of the argument as possible is given in English, in the author's belief that game theory contains ideas of great importance which should be accessible to a wider audience than mathematicians alone, and that, furthermore, it is often not until an argument is expressed in words that one is forced to consider what it really means. There are mathematical passages, but few of these demand more than the capacity to look a symbol in the eye and some elementary mathematical skills, and the ones that do (mostly proofs) are so indicated and may be left out.

There is a serious gap in the literature. Baumol's well-known text *Economic Theory and Operations Analysis* has two chapters which cover the main results on cardinal utility and two-person zero-sum games, but only a paragraph or so on all other types of games (which make up about two-thirds of the present book), and no applications to speak of. The game-theoretic treatment of competitive equilibrium, known as 'core theory', is to be found in one or two modern microeconomics texts, but they present it as an advanced topic; in fact, it is not an especially difficult branch of economic theory. It appears that nowhere are game theory as a whole and its economic applications brought together.

By recognising uncertainty and interdependence as essential features of decision-making, game theory a generation ago laid down a challenge to orthodox microeconomics and seemed to call for a radically new theory. This is no doubt one reason for students' increasing interest in game theory. It may also suggest a subconscious motive for orthodox economists' tendency for many years to dismiss game theory as 'promising but unsuccessful'.

Paradoxically, orthodox neoclassicals have recently 'adopted' game theory. In fact, it is only the least subversive part that they have taken seriously – the theory of cooperative games, in which uncertainty plays a minimal role. This theory has been the basis for major advances in the theory of competitive equilibrium.

Thus for diverse reasons curiosity about game theory has grown; it is on everyone's lips. Yet few know game theory; these trends have resulted only in more frequent but still scattered allusions, footnotes and appendices in the more recent economics and social science textbooks. Hence this book.

Double asterisks ** are used in the text to mark off the more demanding mathematical passages. (Even these, however, need only standard' mathematics for economists'.) The symbol □ indicates the end of a proof. Exercises for the reader are integrated in the text, where they appear enclosed in square brackets.

I would like to thank Elizabeth Fricker for making many valuable suggestions for improving the text, as well as all the other students, colleagues and friends who have given me advice, and the secretaries of Christ Church for making order out of chaos.

Oxford
January 1976

Michael Bacharach

1 Introduction: The Theory of Games and Economic Theory

1.1 THE SCOPE OF THE THEORY OF GAMES

'It is as if someone were to say: "A game consists in moving objects about on a surface according to certain rules . . ." – and we replied: You seem to be thinking of board games, but there are others.' Wittgenstein's reproach [29] could as well be levelled at the game-theorist as at this simpleton who thinks that all games are something like Snakes and Ladders. For the game-theorist, too, recognises only a special kind of game.

In one way the games in game theory are rich in variety, for there is no restriction at all as to their external characteristics. They may be played on a table, by telephone, in the board room, in bed, for any stakes, by any number of players. The definition of the theory is restrictive in quite another way: *the players are perfect*. (Perfect, that is, in reasoning, in mental skills – games in which success depends on dexterity or physical strength have no place in our considerations.) Thus games in game theory are to real games as the ideal consumer's decisions of economic theory are to real consumers' decisions. The theory of games is the theory of *rational* play.

Game theory's conception of games, unlike that of Wittgenstein's straight man, is permissive as well as restrictive. First, a list is made of the 'essential' characteristics of games. Then, *anything* that has these characteristics is called a game. Thus, its methodology is one of abstraction and assimilation. Wage negotiations, price wars, disarmament talks, war itself – all these activities are seen as games. (For the game-theorist, moreover, it is only that which is game-like in these activities that is of interest.)

It may be objected that this is going too far, abstracting too much, confusing the essences of quite disparate things; that along with the word 'game' go phrases like, 'it's only a game', and that this is exactly what can not be said of wage negotiations, or war, so these should not be called games. But it doesn't really matter what term is used. Such implications are not part of the technical game-theoretic meaning of game. To avoid misunderstanding we are sometimes obliged to speak of 'games in the sense of game theory', but this is usually unnecessary.

What does matter is that all games, whatever their outward forms, have a common inner structure. By discovering and understanding this structure, and applying our understanding of it to human activities remote from what we ordinarily call games, we may understand something new about these too. Like all conceptual advances, the theory of games makes one see the most unlikely things in the new way. The economist sees a subtle marginal calculus in the speed at which a man runs for a bus, or in a political party's choice of an election platform; the psychologist sees a defence mechanism in the lighting of a cigarette. In the same way, the theory of games makes one see in the size of a sealed tender, or in an invitation to join an auction ring, or in a call on the hot line from Washington to Moscow, moves in games.

The common structure of all games is the object of investigation of the theory of games. This structure may be expressed, for now, by saying that a game is an activity of two or more human beings whose outcome matters to each of them, and depends in a prescribed way on the actions of all of them.

This raises two problems whose solution constitutes the programme of the theory of games. First, if the outcome of a game depends on everyone's actions, then the most advantageous action for any one player is not well-determined. It will vary according to the choices that the other players make. But the others are in the same boat: their best choices are ill-defined by a symmetrical argument. So there is nowhere to 'begin'.

Secondly, in many games secrecy is of the essence. There is no question of consulting with the other players in order to arrive at a configuration of choices that will produce an outcome that is satisfactory for everyone. Each player could benefit from knowing what actions the others plan to take, but he is constrained by the rules to make his choice in ignorance of them. Games of this kind are called *non-cooperative*. Poker is such a game; war another. Non-cooperative games are doubly problematical. The players have to contend not only with the *interdependence* of their best courses of action, but, because they are obliged to choose *independently*, they have also to contend with *uncertainty*.

Game theory is the theory of what a player should do. Its object is not to explain how people do play games, but to identify individuals' optimal actions in the face of interdependence. It is that part of the theory of rational decision in which the number of rational agents is greater than one and their problems cannot be disentangled.

1.2 ECONOMICS WITHOUT GAME THEORY

It is possible to see much of economic theory as a theory of rational choice. This is most explicit in neoclassical microeconomics. There, *homo economicus* spends his waking hours making cool and calculated decisions designed to bring about the top outcome of a set of possible outcomes which are arranged in a clear order of preference.

This maximising or optimising behaviour is then taken by the neoclassical theorist as a first approximation in describing real behaviour. Such a derivation of 'is' from 'ought' by the use of an ideal type of rational behaviour as approximate description is typical of neoclassical methodology. It has often been made with a hair-raising lack of precaution, but there are circumstances in which it is a quite sensible way to start. We shall refer to this device as the 'normative-to-descriptive step'.

Keynesian macroeconomics appears at first sight to be of an opposite character. It sets out to describe, in aggregated form, the actual reactions of economic agents, however irrational these may be, 'warts and all'. Here we find money illusion, non-adaptive expectations, inflexible consumption habits. Yet even here there are, increasingly, attempts to rationalise the apparently irrational, to explain observed behaviour in terms of latent individual optimising choices. These attempts are made by Keynesians as well as neoclassicals.

Wherever individual optimisation appears – in Keynesian macroeconomics, in the theory of optimum growth, in oligopoly theory – the account that is given of it, or that is simply taken for granted, is that of neoclassical microeconomics. Unfortunately the neoclassical account is grossly deficient. What is amiss is not that it is false but that it deals only with a very special sort of decision problem and probably a very rare one. In a neoclassical decision problem the outcomes in which the agent is interested (a) depend only on his own choices, and (b) depend on his own choices in a known way. It is as if the only game in the world were Solitaire, and all the cards were face up.

The poverty of this theory is one cause of the *idée fixe* of perfect competition in conventional microeconomic theory. In perfect competition, conditions that suit the theory hold good. Consider one individual: first, the outcome of his decision depends on no other individual's decision, for both are too insignificant to affect the options or the prospects of the other; it depends only on the massed decisions of all others. These decisions of others are taken, individually and hence collectively, regardless of our negligible agent. So there is no two-way dependence. Secondly, all the information about these massed decisions that is relevant to the agent's problem is contained in the market prices which they produce – and these prices are assumed to be known to him. So there is no uncertainty. It would be unkind but not unjust to say that a stereotype neoclassical who banishes non-competitive conditions from sight does so because his decision theory cannot deal with them. (This denial of unpleasant facts is a nice example of 'cognitive dissonance' [12] in a theoretical science.)

Three defences can de offered for the concentration on the case of perfect competition in conventional theory.

(i) The first defence is methodological. The heuristic of economic theory, like that of theoretical natural science, is to begin with easy cases. Newton began his theoretical work in celestial dynamics by taking the planets to be point-masses. Having solved the problem in this case he next considered them to be spheres of uniform density, later oblate spheroids of non-uniform density. If we want a description of the steady state, or general equilibrium, of the whole economic system, it is sensible to begin, in the same fashion, with a simplified fictitious case in which we can confidently expect to get answers. Perfect competition is such a case.

(ii) The second defence depends on the fact that economics is a policy science. The demonstrations of Pareto and the Lausanne school, and of their modern successors, that universal competition conduces to an optimal allocation of resources, means that the configuration of variables in a competitive general equilibrium serves as a model for planning.

(iii) The third defence is empirical. It rests on the claim that deviations from competitive conditions are few, or slight, or both.

The second of these defences is a substantial one – but manifestly irrelevant to the adoption of perfect competition as a first approximation or working hypothesis in descriptive economics – to using competition as the take-off point for a normative-to-descriptive step.

The first is fallacious. Competition may indeed be the most

tractable case for existing techniques: but it is also the only tractable case. As soon as one relaxes competitive assumptions, radically different techniques are needed to analyse rational individual decisions. Now for better or worse, explanations and predictions of behaviour in economics are based on what theory discovers to be rational behaviour. Hence these new techniques are needed for description and prediction as well as for defining rational choice in non-competitive conditions. Game theory seeks them directly. If orthodox microeconomics dissociates itself from this search, or if it looks on with an indulgent smile, its appeal to the heuristic of 'increasingly complex cases' is a piece of wishful thinking.

The third defence is implausible. Oligopoly, international trade, external effects in production and consumption, bilateral wage negotiations, cost inflation all manifest one or both of the two 'impurities', interdependence and uncertainty. Are *all* these phenomena to be regarded as 'anomalies', as 'special cases'? What would be left? These, not atomistic competition free of uncertainty and of all externalities, *are* what surround us. The two 'impurities' are no mere trace elements, they are the stuff of economic decisions.

1.3 THE SO-CALLED FAILURE OF GAME THEORY

Some games have solutions and some do not. To be more precise, there may or may not exist choices of actions by the various players that satisfy certain criteria which the theory takes to characterise 'rationality'. In cases where there are solutions, there may be an embarrassment of riches: many different choices may satisfy the solution criteria, so that there is a problem of indeterminacy or 'non-uniqueness'. If it were hoped that the solution of a game would provide a definite answer, once and for all, to some unanswered problem in a human science which has the same form as the game, then both non-existence and non-uniqueness of the game solution dash that hope. The file on the problem cannot be closed.

But the solution of a decision problem is not like the solution of a murder case. If there has been a murder there must have been a murderer, and failure to identify him is a failure to do something that is in principle possible. But just because there is a decision to be made it does not follow that there is a rational solution to the decision problem. The failure of game theory to give unambiguous solutions in certain classes of games does not necessarily imply that the theory is flawed, or inadequately developed. It may be in the nature of things.

In this book we shall meet ordinary games, and problems isomorphic to them in economics and other spheres of human interaction, which give one a strong sense of a kind of intrinsic insolubility. In others one senses that fundamental developments in other disciplines, such as moral philosophy, are called for if determinate answers are to be found. Indeed we cannot exclude *a priori* the possibility that, just as a number of sleeping-car passengers may equally be the murderers of a single fellow-passenger, so all the choices of some set may be just as 'rational' as each other and only distinguishable on grounds of another kind.

It is also possible that the theory has selected the wrong criteria for deciding what is to count as 'rational', and thus as a solution. We shall discuss this question, but the reader must judge for himself in the end. What constitutes rational choice is evidently a question *a priori*; it belongs to philosophy. It certainly cannot be answered by game theory itself, whose results are arrived at by deductive arguments starting from criteria of rationality which have the status of postulates.

1.4 THE STRUCTURE OF GAME THEORY

In the rest of this introductory chapter we give a preliminary sketch of the theory of games, introducing the main categories of games, showing how they relate broadly to each other and indicating which chapters deal with which. The book is organised according to the natural logical development of game theory rather than economics; economic problems are treated as applications of the theories of different types of games. In this section, to avoid interrupting the flow of the game-theoretic argument, we do no more than mention the associated economic problems in passing. It goes without saying that at this stage our preview of game theory can be no more than a series of clips. Many questions are begged, and much must be taken on trust for the time being.

In the simplest kind of decision only one agent affects the outcome. There is no game. The single decision-maker has to reckon only with a passive environment. But even so he rarely knows for certain the consequences of his actions, because this environment usually contains random processes which affect it unpredictably. If these random processes are regular and observable, the decision-maker can attach numerical probabilities to the consequences of his actions. If, for example, his problem is to decide whether to take a raincoat with him on a business trip, he can attach a numerical probability to its raining while he is away, and hence to his getting wet as a consequence of not taking an umbrella.

In games there is an additional – and as we shall see an even more problematical – source of uncertainty for the agent in the free and conscious choices of other players, who can affect the outcome and who may conceal their intentions from him. But this additional source of ignorance does not *replace* ignorance due to randomness of a probabilistic kind. The latter may also be present in games, as it obviously is in 'games of chance'. So a prerequisite for an adequate theory of games is a theory of rational decisions when the outcomes are probabilistic, a theory of decisions 'under risk'.

The seminal book in game theory, von Neumann and Morgenstern's *The Theory of Games and Economic Behavior* [28], begins with a theory of decision under risk. This theory is called the theory of 'utility', though it has little to do with classical utilitarianism. It is so called because von Neumann and Morgenstern establish that if choices under risk are made rationally, they are made just *as if* the agent assigned numerical utilities to the different possible outcomes, then chose the action whose 'expected' utility was highest. Neither the exact meaning of 'expected', nor the criteria of rationality suggested by von Neumann and Morgenstern need concern us at this point.

The von Neumann–Morgenstern theory of utility has become a basic tool in many human sciences. It will be observed that it sits relatively happily in orthodox microeconomics because although it deals with a state of imperfect knowledge its employment in no way depends on the presence of interacting agents. This theory is the subject of Chapter 2.

Chapter 3 describes *zero-sum* games for two players. This part of the theory of games is relatively well know and is counted even by sceptics as a case of unqualified 'success'. Its fame rests largely on the decisiveness and elegance of its theorems, which establish the existence of, and characterise, a unique solution of every such game.

It is however the least useful part of game theory. This is because the interests of the two players have to be very precisely opposed for a game to be 'zero-sum'. There must be no possible outcome which *both* would prefer, however slightly, to some other possible outcome. In other words, for every pair of outcomes if one player prefers the first the other must prefer the second. We show in Chapter 3 that even in a game whose outcome is a division of a fixed sum of money between the two players, the two players' interests are not generally opposed in this diametric way if there are chance elements in the game and if the players have 'risk aversion'. The phenomenon of risk aversion, which is diminishing marginal utility under another name, will be discussed at the end of Chapter 2.

In Chapter 4, still keeping to just two players, we drop the zero-sum condition. That is, we no longer assume that the players' interests meet exactly head-on. They are, as it were, obliquely rather than diametrically opposed: there are at least some prospects that *both* prefer to some others and there may even be some prospect that both prefer to all others. There is thus some measure of common interest, small or great. We now need to know, as we did not in zero-sum conditions, whether or not the players are permitted to cooperate by the rules. However slight the common interest of the players may be, this common interest provides a motive for common action. So according to whether the framing of common strategies is or is not allowed by the rules, the outcome of the game is likely to be very different. It should be noticed that all that 'cooperation' means here is choosing a joint plan of action, deciding together what actions each will take. With cooperation a new choice-making entity comes into being – the *pair* of players – and this new unit picks out a *pair* of courses of action. Cooperation in the sense of the theory of games in no way implies that either player sacrifices his interests for the sake of the other, only that each communicates and co-ordinates with a view to furthering his own unchanged interests by so doing.

If rational players cooperated, they would clearly never choose actions calculated to produce one of the outcomes that both actively disprefer to some other. But in Chapter 4 it is assumed that obstacles of one kind or another prevent them from cooperating. The effect of this proves to be disastrous. Forced to make their decisions in isolation, the players may now choose actions whose expected consequences are undesirable for them both! And let it be clear that these mutually harmful choices are no mistakes – they are dictated by principles of individual rationality. The classic example is the famous game called the Prisoner's Dilemma, which we discuss at length. Also in Chapter 4 the century-old economic puzzle of *oligopoly*, and problems involving *external effects*, are analysed as two-person nonzero-sum non-cooperative games, and the game solution is compared with textbook solutions. The plausibility of the axioms of individual rationality which define the game solution is subjected to scrutiny.

The lesson of Chapter 4 is quite simple. Where there is *scope* for cooperation there is in general a *need* for cooperation. Independent decision-making is liable to end in tears. The conclusion would be banal, but that neoclassical economics and the liberal political philosophy in which it is rooted seem to teach otherwise – that direct negotiation is superfluous, that *private* decision-making leads to everyone's good. This doctrine in economics, as we may by now suspect, depends critically on assuming a regime of perfect competition, one in which there are not two, nor several, but a great multitude of decision-makers. (In Chapter 7 we shall examine this evidently special case and find that game-theoretic arguments lend support to the neoclassical claim – in these neoclassical conditions.)

We now turn to cooperative games, beginning, in Chapter 5, with two-person ones. A cooperative game is one in which the players *can* cooperate – that is, there is nothing to prevent them from coming to an agreement as to what each of them will do. But neither is there anything that obliges them to. They may well not. In a cooperative zero-sum game, for example, there would be no conceivable point even in entering talks: for the players have nothing in common and therefore nothing on which they could agree. Generally, if an agreement is to be 'on' for rational players, it has to pass two tests. (i) It cannot be bettered, from the point of view of both, by some other agreement. (It is this condition that wards off the misunderstanding which leads to tragedy in the parable of the Prisoner's Dilemma.) (ii) It cannot be bettered, from the point of view of one, by going his own way.

Because of (ii), a way is now needed to assess the strength of each player on his 'own resources'. The answer is neat: it is whatever he could expect to achieve if the game he is playing cooperatively were played non-cooperatively. That is, it is given by the solution of the associated non-cooperative game. For this reason, a non-cooperative problem has always to be solved in solving a cooperative one, so that non-cooperative theory as a whole is analytically prior to cooperative theory. Whether cooperation will be consummated, that is whether a pair, or larger group, will end by acting as a unified agent, depends on decisions made entirely non-cooperatively by the elementary decision-units, individuals. In this sense game theory, including cooperative game theory, is 'methodologically individualistic'.

In many cooperative games an agreement that benefits one player but not another would become advantageous for both if the directly benefited player undertook in advance to transfer part of his gains in some form to the other after the game. Such subsequent transfers, considerations, compensations, kickbacks, are known in the literature as *side-payments*. They play an important role in much cooperative theory, and a central one in the *n*-person theory of Chapters 6 and 7.

The main trouble in cooperative theory is indeterminacy. There may be lots of agreements that pass both tests (i) and (ii). The set of all the agreements that do is called the von Neumann–Morgenstern solution set of the cooperative game. There is no obvious principle of rationality by which the two players can pick out a single one of these. This problem is affine to the problem in welfare economics of discriminating between different allocations of resources all of which are 'Pareto-optimal'.

Nash's proposal for getting rid of indeterminacy is original and ingenious, though there is a hint of legerdemain. Nash begins with an analysis of *bargains* as a special class of two-person cooperative games. The essence of a bargaining game is that the outcome of a failure to agree, i.e. a failure to come to terms, is predetermined – it is simply being stuck with what you had. Nash shows that certain postulates uniquely determine a solution which may be thought of as an 'arbitration' from above that would appeal to reasonable men. We apply Nash's analysis to a not unrealistic model of wage bargaining between a single employer and a single trade union. The settlement which the theory uniquely fixes depends on the firm's fixed costs, the general level of wages in the economy and the attitudes to risk of the two sides – in short, on the factors that determine their relative 'bargaining strength'.

Nash then turns to the general two-person cooperative game. Here each player disposes of alternative sanctions that he may take if negotiations fail. So the outcome of a failure is no longer predetermined. Each player must choose the action he would take in such a case. His announcement of his choice constitutes a *threat*. Nash now attempts to solve the indeterminacy problem and define a unique solution for the general or multi-threat cooperative game by the stratagem of breaking it down into a set of simple bargaining games. The description of this attempt ends Chapter 5.

Chapters 6 and 7 are about cooperative games for any number of players or *n-person cooperative games*. In non-cooperative games the step from two to more players, briefly considered in Chapter 4, introduces no new fundamental questions, but here it does. Consider any group of players or *coalition*. As in a two-person cooperative game each person in the group must decide whether cooperating with the others in the group promises him something better than he can get on his own. Now, in addition to this, he must consider whether it promises more than he could get by throwing in his lot with *others* instead, by joining an alternative coalition. The theory assumes that there are no restrictions to inhibit such moves. Anyone can make overtures to anyone, there are no prior commitments, nor acquired allegiances, negotiation is costless, time is unlimited.

It cannot be a foregone conclusion that *any* provisional arrangements will stand up to such relentless, kaleidoscopic machinations. The deals that can do so constitute the *core* of the game. The core is just the n-person generalisation of the von Neumann–Morgenstern solution set. It is the set of agreements that pass both tests (i) and (ii) above and, in addition, cannot be undermined by some new coalition that can do better for itself independently. In n-person cooperative theory the main undertaking is the description of the cores of games. The problem of resolving indeterminacy in cases in which they contain many different agreements is much the same as in the two-person case, and receives little attention.

It may be surmised that if there are many players there is a greater threat to the viability of any set of provisional deals, as there are more regroupings that could overthrow it. Indeed, sometimes the core is 'empty' – there is no set of arrangements that will not be broken by some defection: there is, in short, no 'solution' to the game. A simple situation in which this is so is the sharing of a shilling among three people by majority vote! In such cases the night-long lobbying and counter-lobbying which the theory depicts end in nothing but frustration.

Chapter 6 applies this model of multilateral negotiation to some simple problems in which one or more people have houses for sale and one or more are in the market to buy. Chapter 7 is devoted to a single major problem in microeconomic theory which generalises the simple models of Chapter 6. We describe this problem in the next section.

1.5 GAME THEORY AND COMPETITION

In the model of decision-making that lies at the heart of received microeconomics a single agent 'plays' against a passive environment. As we have suggested, this model of decision works in perfect competition because there prices are 'given'. But why should they be taken as given? True, a single agent among very many equipollent ones has negligible influence, But why should he take his isolated and helpless status lying down? If he joined forces with others, the *cartel* would not be negligible. The traditional analysis of competition that comes from Walras and Pareto side-steps this question when it simply assumes prices to be exogenous or already-determined elements in the decision problem of every individual. The assertion that everyone will bow before the might of market forces is either a question-begging assumption or, at best, an intuitive conjecture of the answer to the question. Game theory shows by a rigorous and compelling argument that such a conjecture would be correct.

The question was asked and answered by a third great neoclassical economist, Edgeworth, as long ago as 1881 [9], using in effect the theory of n-person cooperative games. Edgeworth's 'recontracting' is the supercession of coalition structures we have just been describing. This analysis does not, it will be noticed, involve what are perhaps the most novel and characteristic ideas of game theory, those which involve uncertainty simultaneously with interdependence – those of non-cooperative theory. Wherever there is cooperation there is no uncertainty: what others will do if you do so-and-so is at every turn made very clear to you.

In his celebrated *limit theorem* Edgeworth showed that as the number of people grows in a market in which goods are exchanged, one set of terms-of-trade after another is ruled out – the core 'shrinks'. Finally, only one possible set of exchanges survives. It is one in which there are no cartels. It is the 'competitive equilibrium', in which whoever trades goods with whom, the terms-of-trade are always the same. Furthermore, there is just one set of terms-of-trade that does the trick. And these are none other than the equilibrium relative prices of Walrasian perfect competition. As n 'tends to infinity', negotiation does become superfluous – for any attempt to combine with others in order to trade on other terms than these will be scotched by a defection of one or more of these others to some counter-combine. Deals in the limit are like no deals at all.

The preview we have given will appear to be a meandering story, and so it is. Methodologically, game theory is a branch of mathematics – although its mathematical tools are extremely simple and its arguments can always be seen intuitively. It is not directed to a single question about reality. The object is not to strengthen national security, or to perfect a theory of general economic equilibrium, or a system for winning at poker or oligopoly – though game theory has furthered all these enterprises. It is guided by its own logic. All that is given is its terms of reference – rational decision-making by a number of rational beings. Its structure takes the form of a taxonomy of multi-person decisions: zero-sum or not zero-sum; two-person or many-person; cooperative or non-cooperative. The reader will by now have seen something of the quite different issues that are raised in these different classes of games. Now we are ready to begin.

2 Utility

2.1 RISK AND UNCERTAINTY

The theory of games is part of *decision theory*, the theory of rational decision-making. In this theory a man is said to have a *decision problem* if he has to choose one action out of a number of possible ones with a view to its consequences. The theory assumes that he prefers some consequences to others and sees him as trying to identify the action that leads to the most preferred consequence.

If the decision-maker knows the consequences of each of his possible actions he is said to have a decision problem *under certainty*. Most of the choices considered in economic theory are of this kind. If he does not, his decision problem is one under 'imperfect knowledge'. Most choices in the real world are of *this* kind.

In the latter case, the decision-maker may have ideas about which of the possible consequences of his actions are more, which less likely. Specifically, he may attach numerical *probabilities* to the different possible consequences. If he does so, we shall say – following the terminology of Luce and Raiffa [18] – that he makes his decision *under risk*. Notice that, in this definition, it makes no difference whether his assignment of probabilities is based on empirical evidence, or indeed whether it is justifiable in any way. All that matters for the definition to be satisfied is that he has probabilities in mind and that he confronts his problem using these subjective probabilities.

In game theory, however, any probability assignments that a player may use are assumed to be rationally based – for game theory studies the decisions of perfectly rational agents. (*How* one should rationally assign probabilities either on the basis of experience or otherwise is a large question in its own right. Game theory scarcely touches this question. For a good introduction, see [11].)

If the decision-maker does not know the consequences of his actions, and does *not* attach probabilities to these imperfectly known consequences, we shall say that he makes his decision *under uncertainty*. Uncertainty is, therefore, defined as a state of unprobabilised imperfect knowledge.

There are thus three species of decision problem: under certainty, risk and uncertainty.

Consider two examples of decision-making under risk. (i) A man at a reputable casino is deciding whether or not to bet on black at roulette. (ii) A businessman has to choose between two investments. Which will be more profitable turns on whether there will be war in the Middle East within a year. Of this he thinks there is one chance in four.

In (i), the gambler has, among other things, observable relative frequencies as a basis for his probability assignments and the risk is in this sense objective. Knight [16] reserved the term 'risk' for cases of this sort. In (ii), the businessman does not have relative frequencies to appeal to. The event to which he gives probability $\frac{1}{4}$ is a one-off event; there is no series of 1976s which he can examine to see how often the event happened and how often not. In our terminology, however, *both* (i) and (ii) are cases of decision problems 'under risk'.

Games, in the sense of the theory of games, are decision problems for more than one person. In these decision problems the outcome of my action may (and usually does) depend on your action. In many games I do not know what action you intend and indeed you take pains to see that I do not. I must therefore decide my moves under imperfect knowledge. But what kind of imperfect knowledge do we have here – risk, or uncertainty?

In the real world, the experience of playing with you repeatedly is bound to teach me a good deal about what you are more likely to do – your favourite gambits, your characteristic mistakes – and so give me a basis for probability assignments. But in the theory of games a player does not do this. This is because the theory is a normative one, which sets out to describe rational choices by the players, and it is held that such probabilisation would not be possible if both players are rational.

The argument runs as follows. The fictional opponent of the theory does not make mistakes, nor does he favour particular styles of play which could be recognised and exploited. These things would be understandable, but not rational. The first player therefore has only one means of discovering his opponent's intentions – by arguing *a priori* what it is rational for his opponent to do. But there are different possible conclusions to this *a priori* argument: there are alternative plausible axioms of rational choice. One source of difficulty is that what it is rational for the opponent to do depends on, *inter alia*, what the opponent can discover by the same kind of reasoning about the intentions of the first. (There is, as it were, putting oneself in his shoes, and casting oneself as him putting himself in one's own shoes, and so on. This creates subtle problems for the definition of rational choice, which we shall consider later at some length.)

Whatever principles of rationality the first player appeals to in hypothesising his opponent's intentions, these principles are conceived *a priori*. Hence there are certainly no grounds in experience for the first player to quote numerical odds on different possible principles. What grounds of any kind could there be? Surely none. If this is accepted, it follows that the first player neither knows what principle the other is using, nor can he rationally express his ignorance in numerical probabilities. The same goes, therefore, for the choices of the opponent – choices dictated by the unknown principle. We conclude that games of the kind we have described are played under *uncertainty* by each of the players: there is no way for a fully rational player to assign probabilities to the intentions of a fully rational opponent.

Nevertheless, games may possess elements of risk, side by side with this uncertainty. For one thing, they may contain 'chance moves', such as drawings from shuffled packs or throws of dice which have an effect on their outcomes. In a game of this sort, even if I knew your decision I could only know the consequences of mine 'up to a probability distribution'. Secondly, in von Neumann and Morgenstern's theory it is argued that it may sometimes be advantageous for a player to 'toss a coin' to choose between certain actions. The claims advanced for these 'randomised strategies' are controversial (see Chapter 3); the point here is that if you decided to use a randomised strategy then, once again, even if I knew your decision the outcome of *my* choice would be probabilistic, not certain: I would have to take my decision under risk.

Because outcomes can be probabilistic – synonymously, 'risky' or 'stochastic' – the theory of games could not get far without a theory of choice under risk (although the presence of uncertainty means that this will not be enough).

The kernel of von Neumann and Morgenstern's theory of choice under risk is contained in the answer which they give to the following question. Given the desirability for you of the prizes x, y, what is the desirability of the 'risky prospect' in which you have a $\frac{1}{4}$ chance of getting x and a $\frac{3}{4}$ chance of getting y? A 'risky prospect' of this kind will be written as $[\frac{1}{4}x, \frac{3}{4}y]$ – more generally $[p\,x, (1-p)y]$, where p and $1-p$ are the probabilities of getting the prizes x and y, respectively. Risky prospects are also called *lotteries* and symbols like L, L' will be used to denote them. The outcomes whose probabilities are specified in these lotteries, such as the prizes x, y in the above example, are called *sure prospects.* We have seen that a prerequisite for game theory is an inclusive theory of choice in which the objects of choice may be either sure prospects or lotteries over these sure prospects. This is what von Neumann and Morgenstern provide in chapter 3 of *The Theory of Games and Economic Behavior.*

2.2 UTILITY

In the theory of consumer choice under certainty, the consumer is said to 'maximise his utility'. But saying this carries no implication that there is a utility 'stuff' of which he is trying to 'have' as much as possible. Indeed, in the indifference-curve formulation of Hicks and Allen which has become standard, the theory can be developed without so much as mentioning the word utility, much less asserting the existence of any corresponding substance. In those versions which are expressed in terms of 'utility' the latter is ascribed only a nominal existence. The consumer is said to behave *as if* there were a substance of which he wanted to get as much as possible. Or, we can avoid even metaphorical talk of motives, and merely say that the choice bearing the highest utility index is the choice that will be made.

It is in a similar sense that, in the von Neumann and Morgenstern (VNM) theory of choice under risk, the decision-maker 'maximises his expected utility'. The theory says that the lottery bearing the highest *expected utility* number is the one that will be chosen. We now define the expected utility of a lottery.

Let the outcome of an act α be the prize w with probability p and the prize x with probability $(1-p)$. By choosing α the agent is then choosing the risky prospect or lottery $[pw, (1-p)x] = L$, say.

Suppose we (somehow) attach utility numbers u_w, u_x to the sure prospects w, x. Then the 'expected value' of utility or simply the *expected utility* given by the lottery $L = [pw, (1-p)x]$ is defined to be $pu_w + (1-p)u_x$. It is a weighted average of the utilities of the constituent sure prospects, the weights being their probabilities. (We shall also speak of the expected utility of an act. By this we shall mean the expected utility of the lottery to which the act gives rise.) Notice that the 'expected utility' of a lottery does *not* mean the most likely value of utility to result from accepting the lottery. In most cases, indeed, it is a value which can never transpire. Consider a lottery which yields outcomes whose utilities are 0 and 1 with equal probabilities. The expected value of utility in this lottery is $\frac{1}{2}$, while the utility that eventualises must be either 0 or 1. We see that an 'expected value' is a mathematical construction from possible data, rather than a possible datum.

There is no restriction to dichotomous lotteries. Generally, we may consider the lottery $[p_1x_1, p_2x_2, \ldots, p_nx_n]$, where $x_1, x_2, \ldots, x_n$ are sure prospects, $p_1, p_2, \ldots, p_n \geqslant 0$ and $p_1+p_2+\ldots+p_n = 1$. This lottery has expected utility $p_1u_{x_1}+p_2u_{x_2}+\ldots+p_nu_{x_n}$, or $\sum_{i=1}^{n} p_iu_{x_i}$.

Remember, w, x stand for complete specifications of all relevant consequences of α: in particular, if α is intrinsically undesirable, as in the case of a 'foul means' or an electorally unpopular economic policy, these features are included in the specification of both w and of x and so are registered in the sure-prospect utilities u_w, u_x.

Just as α gives the lottery $[pw, (1-p)x]$, let another act α' give the lottery $[qy, (1-q)z]$. A man who maximises his expected utility, and whose utilities for the prizes w, x, y, z are u_w, u_x, u_y, u_z, chooses α and rejects α' if $pu_w+(1-p)u_x > qu_y+(1-q)u_z$. This choice is written $\alpha \succ \alpha'$, and we say that the man *prefers* α to α'. For it is assumed that what is chosen is what is preferred: effectively, von Neumann and Morgenstern's theory draws no distinction between preference and choice.

Similarly, if $qu_y+(1-q)u_z > pu_w+(1-p)u_x$ then the man prefers (and chooses) α', and we write $\alpha' \succ \alpha$. If $pu_w+(1-p)u_x = qu_y+(1-q)u_z$, he neither prefers α to α' nor α' to α. He is then said to be *indifferent* between α and α', and we write $\alpha \sim \alpha'$. In this case the von Neumann–Morgenstern theory leaves his choice indeterminate.

Suppose that all we know of the expected utilities of the two acts is that one is at least as great as the other, say $pu_w+(1-p)u_x \geqslant qu_y+(1-q)u_z$. Then it follows from the above that an expected-utility maximiser *either* prefers α to α' *or* is indifferent: that is, either $\alpha \succ \alpha'$, or $\alpha \sim \alpha'$. This situation is written $\alpha \succsim \alpha'$. It is sometimes expressed by saying that the man 'weakly prefers' α to α'.

[Verify that $\alpha \succsim \alpha'$ together with $\alpha' \succsim \alpha$ implies that $\alpha \sim \alpha'$.]

A little more notation. $\alpha \succ \alpha'$ may alternatively be written as $\alpha' \prec \alpha$; similarly, $\alpha \succsim \alpha'$ may be written instead as $\alpha' \precsim \alpha$.

It is intuitively obvious, but strictly speaking is an extra assumption about the preference relation $\succ$, that if $\alpha \succ \alpha'$ then it is not true that $\alpha' \succ \alpha$. Equally, if $\alpha \sim \alpha'$, then it *is* true that $\alpha' \sim \alpha$.

The meaning of expected-utility maximisation may be succinctly expressed in a single formula which covers the cases of inequalities in both directions and of equality between the two expected utilities:

$$\alpha \succsim \alpha' \quad \text{if and only if} \quad pu_w + (1-p)u_x \geqslant qu_y + (1-q)u_z. \quad (2.1)$$

[**Prove this.**] Observe that expected-utility maximisation includes ordinary utility maximisation as a special case. Suppose, for instance, that act α gives the consequence w for sure and that α' gives y for sure. We may think of α as yielding the 'degenerate' lottery $[pw, (1-p)x]$ with $p = 1$, and similarly of α' as yielding $[qy, (1-q)z)]$ in which $q = 1$. But then according to equation 2.1, α is (weakly) preferred to α' if and only if $u_w \geqslant u_y$, i.e. the simple utility of w at least equals that of y.

The proposition that the agent 'maximises his expected utility' involves both an assignment (by us) of utilities to the prizes, and the above equivalences connecting $\succ$, $\sim$ and $\succsim$ with $>$, $=$ and $\geqslant$, respectively. The von Neumann–Morgenstern theory arrives at this 'expected utility proposition' (EUP) as its conclusion: the sure-prospect utilities *can* be assigned and the agent *does* maximise his expected utility in the sense of this equivalence. The substance of VNM's claim is that EUP follows if the agent's preferences among his prospects, both sure and risky, are 'rational'.

In sections 2.3 and 2.4 we shall give a definition of rationality in decision-making under risk, and prove that this conclusion indeed follows.

So far we have seen no reason at all why we should accept EUP. But suppose we found we could – what would be gained? According to the expected utility theory preferences are exactly captured in certain numerical magnitudes, namely the expected utilities. Then if A and B are playing a game, the desirability, from A's standpoint, of his choosing an action α and B's choosing an action β, could be expressed in a very simple way, by a single number – A's expected utility of the consequences of these choices. This would be so even should the game that A and B are playing be a game of chance, so that the consequences of their actions α and β can only be known probabilistically. EUP, then, would come in very handy for the theory of games.

In the theory of unrisky choice – the received theory of the consumer – if one utility measure represents preferences, so do others. In fact if u is a utility index for the agent, so is any 'positive monotonic transformation' (i.e. increasing function) of u (e.g. u^3). But $-u$, say, is not (negative monotonic, decreasing), nor is $u-0{\cdot}1u^2$ (non-monotonic, neither always increasing nor always decreasing). Thus, an appropriate utility is partially, but only partially, determined. How well-determined will the utility index for risky choices be if we accept EUP?

The answer is that it is better determined than in unrisky choice: it is determined 'up to a positive linear transformation'. We first illustrate the point, then discuss the underlying reasons why things should be like this.

Suppose that, for our agent,

$$w \prec x \prec y \prec z, \tag{2.2}$$

and also

$$[\tfrac{1}{4}w,\ \tfrac{3}{4}y] \sim [\tfrac{3}{4}x,\ \tfrac{1}{4}z]. \tag{2.3}$$

Try the following assignment of utilities:

$$\begin{array}{ccccc} & w & x & y & z \\ u: & 1 & 2 & 3 & 4 \end{array} \tag{2.4}$$

As we have remarked, the *expected utility* of a sure prospect, such as w, is simply its utility, here 1; for we may think of the sure prospect w as the 'degenerate' lottery $[1.w,\ 0.x]$, say, which has expected utility $1.u_w+0.u_x = u_w$. In the same way, the expected utilities of the 'lotteries' x, y and z are 2, 3 and 4, respectively. Thus the utility assignment (2.4) does represent the preferences (2.2), when we calculate and compare the expected utilities. The utility assignment also represents (2.3). For, calling the left-hand and right-hand lotteries in (2.3) L, L' and writing $Eu(L)$ for the expected utility of L, we have: $Eu(L) = 2{\cdot}5 = Eu(L')$. [Check this.]

Now try a *positive linear transformation* of u, that is, any linear function of u with a positive coefficient of u. For instance, try the utility index u' given by $u' = 7+3u$. We get the assignments:

$$10 \qquad 13 \qquad 16 \qquad 19$$

(2.2) is still clearly okay. Testing for (2.3), we find that the expected utilities of L, L' under the new assignments, $Eu'(L)$ and $Eu'(L')$, both have the value 58/4. Why does the new assignment still work?

For the moment we simply draw attention to the fact that x is still 'half way from w to y' and y is still 'one-third of the way from w to z', and so forth.

Try an increasing but *nonlinear* transformation of u, say $u'' = u^3$:

$$1 \quad 8 \quad 27 \quad 256$$

Notice that this time fractional distances are not preserved; (2.2) is still okay. But $Eu''(L) = 82/4$, $Eu''(L') = 280/4$. Both u' and u'' still represent the non-risky preferences, but only u' still represents the risky ones.

Why is it that we are here more restricted in our choice of utility indices? Mathematically the answer is clear: our assignments must satisfy an extra restriction, provided by (2.3) – since at this stage of the argument we are granting EUP. But can we give a real interpretation of what has happened?

Generally, the more information a measure conveys the more restrictions there are on the number-assignments.

(i) *Associative* measures are the least informative. Their aim is merely to distinguish objects: in numbering the individuals in a group photograph, or court exhibits, the assignments can be anything (provided only that no two items bear the same number).

(ii) *Ordinal* measures (mineral hardness indices, house numbers, the neoclassical consumer's utilities) require the assignments to be in the same numerical order as the order of the objects in terms of the studied characteristic: knowing that object x has a higher number than object y then informs us that it is harder, further along, preferable.

(iii) Yet more informative are *cardinal* measures. In measuring length, or weight, we need assignments which associate, which order, but which, in addition, allow the prediction of the effects of certain more delicate operations: this prediction is accomplished by means of certain corresponding arithmetic operations on the numbers. In the case of length, for instance, end-to-end juxtapositions correspond to additions; or in the case of weight, splitting an object into two balancing portions corresponds to multiplying by 0·5.

So it is with the sure-prospect utilities of von Neumann and Morgenstern. Here, the 'real operations' on the objects (x, y, . . .) are *probability mixtures*. The result of this mixing is a lottery. A lottery is much like a little heap of divided objects for weighing. Let x be a cheesecake: I offer you successively smaller weight-fractions of the cheesecake if I divide it symmetrically into more and more portions; I offer you successively smaller probability-fractions of the cheesecake if I give you the *whole* cheesecake *on the condition* that the ball stops in narrower and narrower intervals of a roulette wheel's circumference. (If it is a probability-fraction that I offer you, I will never take a knife to the cheesecake: either you get all of it, or you get none at all.) Not all objects may be divided in weight; but all may be divided in probability. Probability-division allows us to fraction arbitrarily an arbitrary object of preference. We may ask 'how much of x?' of any object, animal, mineral, vegetable, divisible or not, marketable or not.

This ability to fraction x in any way we please puts us in a position to ask: *how much desirability* does x have? For all we need to do is to flip this question over into the following form: how much of x (what *probability*) is needed to give a mixture containing in it a certain total desirability? This is the trick to von Neumann and Morgenstern's theory.

To the real operations of forming probability mixtures or lotteries of the xs and ys correspond the arithmetic operations of multiplying their utility measures by fractions equal to their probabilities, then adding: that is, forming expected values of the measures. Our example illustrates that these operations successfully predict the effect of the real operations – equation 2.3 represents just such an effect – but they do so only if the number assignments are chosen restrictedly, just as are weights.[1]

We shall now begin setting out the argument that if the agent is *rational* in his preferences among lotteries, EUP follows – we *can* find a utility assignment u such that all his choices among lotteries are predictable by computing and comparing the utility 'weights' (expected values of utility) of the lotteries. That is, the lotteries preferred by this rational agent are precisely those whose expected utility values are high when we compute these expected utilities using our measure u. (Note well that our discussion of the degree of freedom of choice we have in choosing the measure suggests that if the EUP is true for some index u, then it is *also* true for any other index u' of the form $u' = a + bu$ with $b > 0$. [Prove this.])

As a preliminary we observe that this can only be so if there is a

---

[1] Measures of weight are in fact subject to an extra restriction that does not apply to utilities: the measure zero must be assigned to the (imaginary) object which would not affect the scales, but there is no analogous object of preference. A more exact parallel to utility is therefore 'distance north'.

certain order in his preferences: he certainly cannot be allowed to prefer, as it were, what he pleases. For example, look again at equations 2.2 and 2.3. If EUP holds, it follows that he *must* be indifferent between x for sure, and the lottery $[\frac{1}{2}w, \frac{1}{2}y]$.

2.3 AXIOMS FOR RISKY CHOICE

This seems a bit unreasonable. Surely you are the sole judge of what is preferable for you? But the question is not whether you are entitled to your preferences, but whether they are rational. EUP follows from certain axioms of rational choice under risk by the theorem given in the next section. If then you have preferences (equations 2.2 and 2.3) but do *not* have $x \sim [\frac{1}{2}w, \frac{1}{2}y]$, there are only three possibilities: (i) there is a mistake in the proof of the theorem, (ii) you are irrational, (iii) the 'axioms of rational choice' are wrongly so called. However, although the restrictions implied by EUP are strict, the axioms from which they follow *seem* eminently reasonable, as we now see. The following is a particularly undemanding set, which suffice for the EUP proof of section 2.4. We first give the whole set in a fairly abstract, mathematical form, then make some interpretative comments – and some sceptical ones.

(1) (Transitivity.) For any three lotteries L, L', L'', if $L \succsim L'$ and $L' \succsim L''$ then $L \succsim L''$.

(2) (Desirability of probability of preferred prize, or 'positive association'.) For any sure prospects x, y such that $x \succ y$, $[px, (1-p)y] \succ [qx, (1-q)y]$ if and only if $p > q$.

(3) (Archimedean or continuity postulate.) For any sure prospects x, y, z for which $x \succsim y \succsim z$, there exists a probability p such that $y \sim [px, (1-p)z]$.

(4) (Postulate of compound probabilities.) For any x, y and any probabilities p, p_1, p_2, consider the following 'two-stage' lottery whose prizes are *lotteries* over x, y: $L = [p[p_1x, (1-p_1)y], (1-p)[p_2x, (1-p_2)y]]$. Then $L \sim [rx, (1-r)y]$, where $r = pp_1 + (1-p)p_2$. (That is, the agent *can* evaluate a two-stage lottery, and he *does so* by considering simply the probabilities of the ultimate prizes.)

(5) (Independence, or substitutability of indifferent prospects.) For any prospects k, l, m (sure *or* risky), if $k \sim l$, then $[pk, (1-p)m] \sim [pl, (1-p)m]$.

Axiom (1) is weak enough. It says that *if* the agent makes pairwise comparisons between L and L' and between L' and L'', then he must do so consistently. (2) is really two requirements in one. First, it requires that the agent *can compare* risky prospects. (Specifically, for any two non-indifferent sure prospects x, y, he can compare any two lotteries involving them – and this includes degenerate lotteries, i.e. x, y themselves, since p may be 0 or 1.) Secondly, he prefers more of a good thing to less – more probability, that is. Once we accept the first part the second part seems innocuous. True, it may be hard to discriminate very fine probability differences, but this objection belongs to psychology, not rational choice theory.

Axiom (3) adds little to (2). We comment only on the 'strict' case $x \succ y \succ z$. Consider lotteries between x and z with p, the probability of the prize x, gradually increasing from 0 to 1: their desirability is always increasing, by axiom (2). It must therefore at some point coincide with the level of desirability of y, unless either it jumps at some point (which would seem quirkish), or x, z lotteries are non-comparable with third prospects y. (So a bit more comparability is required here.)

Axioms (4) and (5) both concern lotteries over lotteries. Like (2) and (3), a major part of (4) is that certain types of prospect (two-stage lotteries) are evaluable. [Convince yourself that $r = pp_1 + (1-p)p_2$ is simply the probability of winding up with x.]

Axiom (5) is the most contentious postulate. It says that if any two prospects (sure ones or lotteries) are equivalent whole, then equal probability-fractions of them are equivalent when these appear as components in mixtures. Remark that this – like all the other axioms – holds for weight, or length, when suitably re-worded.

How can axioms like these, descriptions of rationality offered *a priori*, be put to the test? The standard method is by considering 'counter-examples' – choices which themselves look perfectly reasonable *a priori* but which violate one or more axioms: there ensues a trial of strength between the counter-example and the axiom.

Axiom (2) ('positive association') is challenged by supposing x, y to be, respectively, surviving and not surviving a dangerous adventure. If p is some high probability, the daredevil racing driver, bullfighter or explorer may have the preferences

$$[1.x,\ 0.y] \prec [px,\ (1-p)y] \succ [0.x,\ 1.y].$$

This is an apparent violation of (2). But defenders of the theory argue that the risk of the exploit should be considered to be part and parcel of the 'sure prospects': according to this defence, x and y are not 'proper' sure prospects.

Axiom (4) has a well-known counter-example which may also come from mis-specification. In Paris you can play fair wheels of chance whose prizes are tickets in the *Loterie Nationale.* According to (4) there would be no point in this rigmarole; but apparently people enjoy it.

The most telling counter-examples concern axiom (5). Samuelson's [21] challenges it directly. Notice that the indifferent prospects which are supposed to be substitutable may either be lotteries or sure prospects. Samuelson thinks the axiom questionable if one of them *is* sure, because when it appears fractioned in a lottery it loses this quality, to which it owed its essential desirability. Say £200 $\sim$ [$\frac{1}{2}$£1000, $\frac{1}{2}$£0]. Then according to axiom (5), for any sure prospect x and any p, we have

$$[px, (1-p)£200] \sim [px, (1-p)[\tfrac{1}{2}£1000, \tfrac{1}{2}£0]].$$

Putting $x =$ £500, $p = \frac{5}{6}$, we get

$$[\tfrac{5}{6}£500, \tfrac{1}{6}£200] \sim [\tfrac{5}{6}£500, \tfrac{1}{6}[\tfrac{1}{2}£1000, \tfrac{1}{2}£0]] = [\tfrac{1}{12}£1000, \tfrac{5}{6}£500, \tfrac{1}{12}£0].$$

But many would prefer the right-hand lottery: £200 contingent on 'throwing a six' is not really worth the sacrifice of the chance (not all that much smaller) of the really rich prize.

The final counter-example is the famous Allais paradox [1]. Like the last example this involves embedding a comparison of prospects in an elaborated lottery. However, it does not challenge axiom (5) unequivocally, because to show the impermissibility of the choice Allais claims to be natural you have to use all the axioms (1) – (5). Let w, x, b be sure prospects such that $w \prec x \prec b$ (w for 'worst', b for 'best'), and let L be $[rb, (1-r)w]$ for some r, $0<r<1$. Suppose that

$$x \succ [px, (1-p)L] \quad \text{for some } p,\ 0<p<1. \tag{2.5}$$

Then it can be shown that axioms (2)–(5) entail that

$$[pw, (1-p)x] \succ [pw, (1-p)L]. \tag{2.6}$$

We do not give the proof, but one can see intuitively the force of the argument from (2.5) to (2.6). (2.5) may be written

$$[px, (1-p)x] \succ [px, (1-p)L], \tag{2.5$'$}$$

which, it can be shown, entails that $x \succ L$. This in turn makes equation 2.6 obvious enough.

In Allais's example, acts α_1, α_2, α_3 and α_4 give one the following prospects:

α_1: £1000; α_2: [0·01 £0, 0·89 £1000, 0·10 £5000];
α_3: [0·89 £0, 0·11 £1000]; α_4: [0·90 £0, 0·10 £5000].

Most people prefer α_1 to α_2; and α_4 to α_3. In justifying the second preference they argue: 'I can't avoid the risk of £0 and the chances of £0 are very close, so why not go for the much bigger prize?' [Consider what choices you yourself would make.]

Now put $w = £0$, $x = £1000$, $b = £5000$, $r = 10/11$, $p = 89/100$. Then [check this]:

$$\alpha_1: x; \qquad \alpha_2: [px, (1-p)L];$$
$$\alpha_3: [pw, (1-p)x]; \qquad \alpha_4: [pw, (1-p)L].$$

So by (2.5) and (2.6), if α_1 is preferred to α_2, α_3 should be preferred to α_4. Most people's choices are therefore irrational judged by our axioms. Who is right? The decision theorist Savage records [22] that his own intuitive preferences were $\alpha_1 \succ \alpha_2$, $\alpha_4 \succ \alpha_3$, but that when he contemplated the options expressed in the form of (2.5′) and (2.6), he wished to revise them.

If serious men trying to be rational violate the von Neumann–Morgenstern axioms, it is not surprising that they do not work as descriptive psychology. There are lots of ways in which people commonly violate axioms (1)–(5) (see [10]). Among the most picturesque are 'number dazzle', and the disposition of punters to take bets at certain favourite numerical odds.

2.4 THE EXPECTED UTILITY PROPOSITION

We have asserted that if an individual obeys axioms (1)–(5), then he behaves as if he is maximising an expected utility. We now sketch a proof. This proof is only for choices over *dichotomous* lotteries – but the generalisation is merely a bit more complicated and does not involve any new ideas. We assume, too, that there are a worst and a best sure prospect – there are limits to the imaginably good and the imaginably bad. The proof is 'constructive': it proves the existence of the expected utility by actually making a utility assignment which does the trick.

THEOREM (the expected utility proposition). If a decision-maker obeys axioms **(1)–(5)**, then there is a function u of sure prospects such that for any sure prospects w, x, y, z, and any probabilities p, q,

$$[pw, (1-p)x] \gtrsim [qy, (1-q)z] \quad \text{if and only if} \quad pu(w)+(1-p)u(x) \geqslant qu(y)+(1-q)u(z).$$

PROOF. Denote the worst and best sure prospects by w and b, respectively. Consider any dichotomous lottery, say $L = [px, (1-p)y]$, where x, y and p are arbitrary. Now by the Archimedean axiom **(3)

$$x \sim [p_x b, (1-p_x)w] \quad \text{for some } p_x;$$

similarly,

$$y \sim [p_y b, (1-p_y)w] \quad \text{for some } p_y.$$

Now bring in the contentious independence axiom **(5)**, and we have

$$L \sim \big[p[p_x b, +(1-p_x)w], (1-p)[p_y b, (1-p_y)w]\big].$$

Reducing the right-hand side by using the (compounding) axiom **(4)**, it is indifferent to $[rb, (1-r)w]$, where $r = pp_x+(1-p)p_y$.

So using axiom **(1)** (transitivity), we have that

$$L \sim [rb, (1-r)w]. \tag{2.7}$$

In exactly the same way, for *any* other dichotomous lottery $L' = [p'x', (1-p')y']$, where x', y' and p' are arbitrary sure prospects and an arbitrary probability,

$$L' \sim [r'b, (1-r')w], \tag{2.8}$$

where $r' = p'p_x'+(1-p')p_y'$.

Now by axiom **(3)** (positive association), the right-hand lottery in (2.7) is weakly preferred to ($\gtrsim$) the right-hand lottery in (2.8) if and only if $r \geqslant r'$. So by transitivity **(1)**, $L \gtrsim L'$ if and only if $r \geqslant r'$.

Hence, we shall have proved the theorem if we manage to assign utilities to sure prospects in such a way that

$$r \geqslant r' \quad \text{if and only if} \quad Eu(L) \geqslant Eu(L'). \tag{2.9}$$

We now do just this. Make the sure prospect assignments

$$u(w) = 0, \qquad u(b) = 1,$$
$$u(x) = p_x, \qquad u(y) = p_y, \qquad u(x') = p_x', \text{ etc.}$$

Then

$$\begin{aligned} Eu(L) &= pu(x)+(1-p)u(y) \\ &= pp_x+(1-p)p_y \\ &= r. \end{aligned}$$

Similarly, $Eu(L') = r'$. This establishes equation 2.9. □ **

We have, then, displayed a utility index u which makes the EUP hold good for an agent who obeys axioms **(1)–(5)**. u will be called a 'valid' or 'legitimate' utility index for the agent. Remember, if u *is* a valid VNM index, so also is any index $u' = a+bu$, with $b>0$. Another way of saying this is that if an agent obeys the VNM axioms of rational choice, and so has *some* valid VNM index, then he has a valid VNM index of arbitrary origin and scale. Take the case in which the sure prospects are various money prizes x ranging from w to b, and let u be the utility index constructed in the proof we have just seen, so that $u(w) = 0$, $u(b) = 1$. Suppose we would like to have an index (u' say) which assigns zero utility to £0 and unit utility to £1000. u' is valid if it is of the form

$$u'(x) = a+bu(x). \tag{2.10}$$

Now we can find an index u' satisfying equation 2.10 and with values zero and unity for the sure prospects £0 and £1000, as required, if we can find a and b for which

$$0 = a+bu(0), \qquad 1 = a+bu(1000).$$

We can. The a and b we need are just

$$a = \frac{-u(0)}{u(1000)-u(0)}, \qquad b = \frac{1}{u(1000)-u(0)}.$$

2.5 THE UTILITY OF MONEY AND RISK AVERSION

Suppose that the prospects a VNM man is considering are monetary: the sure prospects are receipts, or losses, of amounts of money, nothing else changing in the man's circumstances; the risky prospects are amounts of money to be received or lost with certain probabilities – that is, they are bets. We know from the last section that the man has (as if) utilities for different amounts of money-for-sure. That is, he has a utility function for money. (More exactly, he has a utility function for money gains and losses relative to his current state.)

How do these utilities depend – for surely they must – on his current state? How, in particular, do they depend on his present wealth? There is one assumption about this dependence that is commonly made and which we shall have occasion to use later, in discussing the arbitration of bargains (section 5.6) – the assumption of diminishing marginal utility of money. In this section we give a justification of this assumption – and enter a qualification.

First note two points about actual people's VNM utility indices. The first is that they may not have them – for they may not obey the VNM axioms in making their choices under risk. Thus, any derivations of VNM utility functions from observed behaviour should be regarded sceptically. The second is that the theory pronounces no judgement on a man's preferences, provided that they satisfy axioms **(1)**–**(5)**; nor, therefore, does the theory approve or disapprove the corresponding VNM utility function. For example, VNM utilities that rise less and less with each increment of money ('diminishing marginal utility') are no more and no less 'rational' than ones that increase ever faster, or decline!

We begin with the following very commonly observed feature of choices among sure and risky monetary prospects.

Observation 1. A man prefers £y for sure to the lottery L which gives fifty-fifty chances of £$(y+\delta)$ and £$(y-\delta)$.

Here δ is any positive sum. Formally,

$$y \succ [\tfrac{1}{2}(y+\delta), \tfrac{1}{2}(y-\delta)] = L, \qquad \delta > 0. \tag{2.11}$$

A man with preferences like these backs away from what is an actuarially fair bet. He is said to be *risk-averse*.

Now the expected money value of the lottery or bet L is clearly y. So (2.11) amounts to saying that the man would rather have the expected money value of the bet for sure than a fifty-fifty lottery with the *same* expected money value. If u is a VNM utility function for the man, this may be written

$$u(EL) > Eu(L), \tag{2.12}$$

in view of the fact that $EL = y$.

There is presumably some money amount y^* less than y such that the man is just indifferent between the lottery L and y^* for sure. Then the quantity

$$\frac{y-y^*}{y} = \frac{EL-y^*}{EL} = \rho, \quad \text{say,}$$

is called the *risk premium* of the bet L. Let us call y^* the *certainty equivalent* of the bet L; then the risk premium is the percentage shortfall of the certainty equivalent below the money expectation of the bet.

The term 'premium' is used because, if the man had to take the bet L, he would be just willing to pay the amount ρy or ρEL for insurance cover. The insurance policy would stipulate that if the $y-\delta$ outcome transpired the insurer would pay the insured the amount δ; if $y+\delta$, the insured would pay δ to the insurer. If he bought this policy the man would face $y-\rho y$ for sure. But $y-\rho y = y^*$. Since $y^* \sim L$, the man would be precisely as well off paying ρy for the insurance cover as facing the bet L unprotected.

For example, suppose you had to do something which would leave you £150 better off or £50 worse off than you are now, with probabilities 1/2 and 1/2. Here $EL = y = 50$. If you were just willing to pay £10 for the above type of cover, your risk premium for this lottery would be 0·2. [Check this.]

We now show that risk aversion is equivalent to diminishing marginal VNM utility of money. Figure 2.1 shows money receipts on the horizontal axis and on the vertical axis a VNM utility index for the agent. The heights RR', SS' are the VNM utility indices of $y-\delta$ for sure and $y+\delta$ for sure. SS' is greater as long as the man prefers more money to less.

M is the midpoint of RS, so $OM = y$. MM' must be the arithmetic average of RR' and SS', that is $MM' = \frac{1}{2}RR' + \frac{1}{2}SS' = \frac{1}{2}u(y-\delta) + \frac{1}{2}u(y+\delta)$. That is, $MM' = Eu(L)$.

Now we have seen that risk aversion implies that $u(EL) > Eu(L)$ (2.12). In words, the VNM utility of y for sure exceeds MM'. So the curve showing the VNM utilities of sure money receipts passes through a point such as N in the figure, a point above M'.

If δ is a small quantity, this implies that the *marginal* utility of sure money is diminishing in the neighbourhood of y. The converse – that diminishing marginal VNM utility of money implies risk aversion – is easily demonstrated by retracing the steps in the above argument.

Say the utility function is the curve drawn through R', N and S' in the figure. Then the certainty equivalent y^* of the bet is easily

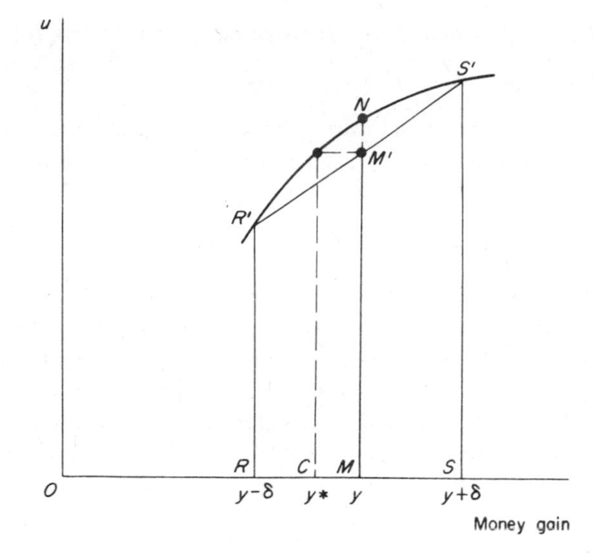

Fig. 2.1 *Risk aversion*

found by drawing in the broken lines. The risk premium ρ can be read off as $CM : OM$.

What might determine the size of the risk premium? Once again, we shall work out the implications of a very common observation.

> *Observation* 2. For given y and δ, ρ is smaller the larger is the agent's wealth.

First, we re-express this empirical 'law' so as to see how ρ varies with y and δ rather than with wealth. Suppose a man with wealth W contemplates the bet L. Let us measure the two possible outcomes of taking the bet in terms of his wealth after the bet rather than his net receipts from the bet. Then $L = [\frac{1}{2}(x-\delta), \frac{1}{2}(x+\delta)]$, where $x = W+y$. Notice the implicit assumption that W and y are of the same stuff, e.g. of the same liquidity.

Observation 2 now reads:

> For given δ, the larger is x the smaller is the risk premium of the bet $[\frac{1}{2}(x-\delta), \frac{1}{2}(x+\delta)]$.

This phenomenon is called *decreasing risk aversion*. It may also be expressed by saying: the risk premium of a bet depends inversely on how much is at stake relative to one's (expected) fortune.

It can be shown that (**if u is twice differentiable**), for small δ,

$$\rho = \tfrac{1}{2}|u''/u'|x(\delta/x)^2.$$

$|u''/u'|$ is a measure of the curvature of the utility function u, and $(\delta/x)^2$ of the variance of the relative deviation of the outcome from its mean x. We may write

$$\rho = \tfrac{1}{2} \times \text{curvature} \times \text{mean} \times \text{relative variance}.$$

Since the relative variance decreases with x, given δ, we would certainly have decreasing risk aversion if the product of curvature and mean were constant as x increases. This product is called the 'degree of relative risk aversion'.[1] It will be constant only if the curvature of the utility function of money tends to zero (flatness) as x tends to infinity. That is, if a rich man has a near linear utility function of smallish absolute changes in his wealth, while a poor man's is sharply curved. In Figure 2.1, N would lie scarcely above M' for a rich man, well above M' for a poor one.

In short, Observation 1 means that the VNM utility function of increases in wealth is concave, and Observation 2 that it gradually flattens out. But not all empirical evidence supports this picture. We have seen that paying a premium for insurance against risk above the 'actuarially fair' price implies risk aversion. In the same way, if a man is prepared to take a bet – to take on a risk – at worse than actuarially fair odds, as he does in a casino, then his implied VNM utility function of money shows 'risk preference' and curves convexly instead of concavely. Some people do do this. Furthermore, Friedman and Savage [13] have pointed out that frequently the same man both buys insurance and habitually takes small gambles at less than actuarially fair odds.

This behaviour is no puzzle if we accept that the gambler is buying not merely a lottery over money outcomes, but as it were the exciting box in which the lottery is packed. If, on the other hand, we find this implausible, the Friedman–Savage observation may also be accounted for by a VNM utility function of money of a different shape. Notice that the stakes in the insured lottery are larger than those in the observed gambles. The utility function drawn in Figure 2.2 then shows that there need be no inconsistency in both paying for insurance and betting at unfavourable odds.

---

[1] More generally, the condition for decreasing risk aversion is that the degree of relative risk aversion does not increase as fast as $(\delta/x)^2$ decreases.

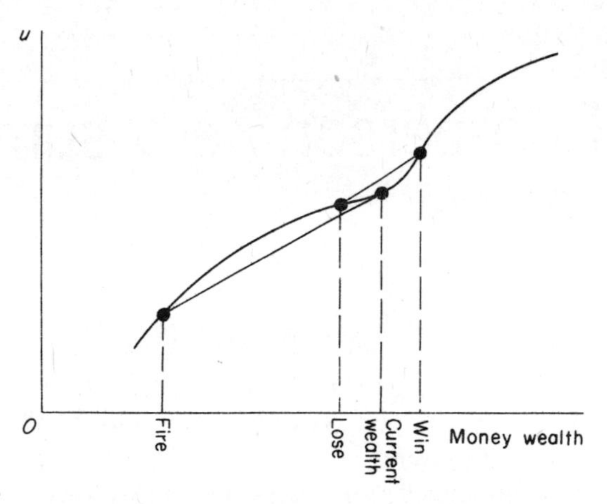

Fig. 2.2 *The Friedman and Savage case*

3 Two-person Zero-sum Games

3.1 THE DEFINITION OF A GAME

Whatever has the following four elements is a game:

Element **(1)**. A well-defined set of possible courses of action for each of a number of players.

Element **(2)**. Well-defined preferences of each player among possible outcomes of the game and, *à la* von Neumann and Morgenstern, among probability distributions or mixtures of its outcomes.

Element **(3)**. Relationships whereby the outcome (or at least a probability distribution for it) is determined by the players' choices of courses of action.

Element **(4)**. Knowledge of all of this by all the players.

Elements (1) and (3) are given in the *rules* of the game and (2) and (4) describe the players.

We have put these four constitutive properties of games both abstractly and concisely. Their meaning will become clear as we begin to look at examples of games. The nearest sort of real-life game to the abstraction defined in elements **(1)**–**(4)** is probably the parlour game – at least in respect of the first three elements. Once the players have agreed to play, what they can and cannot do is clearly laid down in Hoyle – element **(1)**. In that authority one can find, too, the information referred to in element **(3)**: if the various players do this, that, or the other, we can work out who will win; or, in games of chance, we can calculate the odds on one or another player's winning. Element **(2)** simply attributes to the players the kind of consistency in their preferences which we analysed in the last chapter: *if* I prefer £10 to £5, then I also prefer a good chance of £10 rather than £5 to a poor chance of £10 rather than £5; and so on.

In economic life, as in the parlour, activities are to be found that greatly resemble games in the sense of elements **(1)**–**(4)**.

All the defining properties **(1)**–**(4)** are restrictive. No doubt if there were a deviation from any of them we would still, in ordinary language, call what was taking place a game; but it would lie outside the boundaries of game theory. Ordinary words, as Wittgenstein argues [29], have senses that are always adaptable and elusive; game theory deals with an ideal type of game, bounded by hard defining lines. If a real activity deviates in any way from the paradigm described in elements **(1)–(4)**, it is not a 'game', and the theory of games cannot be used directly in its analysis. This severely limits the range of application of the theory.

Element **(1)**, for instance, means that the model is not appropriate at the moment when, before he makes his decision, a player has to engage in 'search activity' to discover what choices he has. But this is often the case: consider the unknown production techniques available to a firm [26], or the unknown market options facing a redundant worker [17]. Element **(2)** disqualifies players who violate the VNM axioms – as most subjects in laboratory experiments do. This is in fact no objection to the theory, which is meant to be normative, if these violations are due merely to human imperfection. What is perhaps an objection to it is that element **(2)** rules out 'getting to know you' effects. Playing with someone (like dancing with a king) may affect what you wish for him and for yourself. Consider the 'game' played out between terrorists and hostages.[1] Interactions of utility functions are not irrational, but there is no room for them in the theory, whose treatment of preferences is thoroughly static. Element **(3)** implies that the outcome of the game cannot be affected by any outside cause, unless it be random: no conscious agent can interfere with it. In particular, no partisan spectator can give moral support: in the conditions required by **(3)**, therefore, 'home' and 'away' games should have the same outcomes. Element **(4)** is asking a great deal. Consider a parlour game like poker. The requirements of elements **(1)–(3)** are well satisfied; but **(4)** is violated unless I know, on top of everything else, your exact degree of risk aversion.

The players do not have to be individuals. They might equally be partnerships, teams, firms – any groupings which can choose and act as units. From the point of view of the theory it does not make any difference, and the terms 'players', 'individuals' and 'persons' are used interchangeably. Such terms as 'two-person game' should therefore be understood as 'two-party game', and so forth.

---

[1] A game in which the players' utilities interacted in an unexpected direction was the kidnapping game played in the basement of the Spaghetti House restaurant in London in September 1975. On the sixth day, when the kidnappers gave themselves up to the police, their Italian hostages threw their arms round their captors and wept.

If there are more than two 'players' in a game (that is, more than two elementary decision-taking units) the rules may allow subsets of them to collaborate – *coalitions* may be formed. The rules will have to be very clear as to whether this is permissible if there is any possible advantage to be gained from collaboration. Indeed, even in two-person games the rules may need to stipulate whether the two players may collaborate, for there certainly are two-person games in which both players may benefit by acting together. Much of the later development of game theory (Chapters 5–7) concerns the rationality of collusion. Games are called *cooperative* or *non-cooperative* according as collusion (collaboration, forming coalitions) is or is not *permitted by the rules*. 'Monopoly' is a non-cooperative game; a round in 'Diplomacy' is a cooperative game. Wage bargaining is a cooperative game; oligopoly, where cartels are illegal, is a non-cooperative game.

In all games, cooperative or non-cooperative, there is interdependence. True, an activity in which the outcome of interest to any player depended only on his own actions would not actually violate our definition of a game in elements **(1)–(4)**, but it would hardly merit calling a game. The interest of games lies in the way the outcome depends on a multiplicity of choices, and the theory of games studies the problem of rational choice when this is so. Game theory would be quite unneccessary to analyse the decision problems faced by a number of people in a room all playing patience.

In a cooperative game the players may collaborate or act together. But what does 'act together' mean? Say there are two players, and label them A and B. Instead of A choosing one out of his possible courses of action and B doing likewise, in a cooperative game A and B can, as a unit, *choose* one out of all their possible *pairs* of courses of action. This requires communication between them. Suppose that some particular pairing of actions is beneficial for both A and B. If they cannot communicate they may none the less by chance, or by guessing what is in each other's mind, make individual choices which together compose this beneficial pair; but they cannot be said to 'choose' the pair.

In the theory of non-cooperative games the main project is to determine an *equilibrium* for players independently choosing courses of action in pursuit of their interests. These choices are assumed to be made with ideal rationality and intelligence: in game theory players do not make mistakes. The notion of 'equilibrium' is closely related to the orthodox economic one: but it is necessarily broader and subtler, for here, because of interdependence, it is impossible to define the maximising choices of individuals one at a time, independently of the choices of the others.

In cooperative games the problem of saying what rational choice is when others' choices are both pertinent and unknown is no longer with us. But new problems arise to replace the old. Payoffs accrue to coalitions, and so before we can say we have solved a cooperative game we must say how the proceeds are to be shared between the coalition's members. This problem in game theory is like a notorious problem of welfare economics: there, it is clear that a solution to the problem of the maximum welfare of society is somewhere on the Pareto frontier – but whereabouts? The problem is also, however, distinctively game-theoretic. For here the final distribution of utility must recognise the strategic strengths of the individuals: it must be sensitive to the fact that, if a particular coalition does not promise me at least what I can get by independent action, I would be a fool to join it. In the theory, fools don't play games.

From now until Chapter 5 we shall confine our attention to *non-cooperative* games.

3.2 MOVES AND STRATEGIES

In elements **(1)** and **(2)** above we referred to the *courses* of action of the players. This was deliberate. The outcome of a game of chess is not decided by the pair of opening moves, nor the outcome of a game of poker by which players stake themselves in. Clearly the 'courses of action' which determine the outcome must be sequences of moves game-long.

In the later stages of a game certain moves are ruled out for me, both by my own previous moves and by the previous moves of the other player or players. I cannot play Queen's bishop to K5 at my fifteenth move if I have not manoeuvred my piece into position, or if it has been captured. Further, if a game is to be a game as we have defined it then at each stage a player must know his permissible moves. The upshot is that the only courses of action that are of interest are these: sequences of moves, each move in such a sequence being a move the player will make *given that* he finds himself with such and such permissible moves. But there is also something else: generally, at each stage the player will know more things relevant to his next move than the set of actions permitted by the rules of the game: for instance, he may know that the man on his left must be, or cannot be, holding such and such a card.

A course of action that prescribes exactly what a player would do in every conjuncture or current state in which he may conceivably discover himself to be is called a *strategy*. It is an extended contingency plan. 'Match a price fall, ignore a price rise' is a strategy for an oligopolist – supposing that price changes are the only moves his rivals will make. In principle, a strategy may be written down in advance and handed to an agent. If everyone did this the outcome of the game would be settled before play began. But this is precisely what element **(3)** says. [Why do people not play in this way?]

Notice that a player does not put himself at a disadvantage by deciding on a strategy and sticking to it. He would do no better by making up his mind as he went along: for his total useful information at the moment when he can no longer put off deciding on his fifteenth move is not made any greater by this procrastination. Although he could not have known the actual situation in which he finds himself, he could have hypothesised it. True, in selecting a strategy he must prepare lots of conditional fifteenth moves which will turn out to have been prepared unneccessarily. The theory neglects the tiresomeness of this chore, just as it neglects the burden on him as he chooses a strategy of holding in mind the multifarious branching of possible developments of the play. For he is assumed (element **(4)** implies this) to have unlimited mental capacities.

Thus any game in *extensive form*, that is any real move-by-move game, may be reduced analytically to an equivalent game in which each player has only one 'move' – choosing a strategy. The game is then said to be in *normal form*.

EXAMPLE: RUSTIC POKER

This game is also known as Bluffing and as Hi-Lo [28, 14]. A hat contains two cards, marked Hi and Lo. Each of the two players, whom we shall call A and B, stakes himself in by putting on the table an 'ante' of a (shillings, DM or whatever). Suppose it is A's lead. He draws one of the two cards and looks at it without letting B see it. He now either 'folds', in which case B pockets his (A's) ante of a, or he 'raises' by $b-a$, thus making his total stake b. If A folds and B takes his ante, the game ends there. If A raises, it is now B's move.

Now B either folds, losing his ante to A, or he 'calls'. In the latter case A receives or pays b according as he has Hi or Lo. The game is now over.

We want to specify A's and B's *strategies*. Observe that when A draws his card, *chance* makes a 'move'. This has the consequence, as we shall see, that some of the payoffs in this game are stochastic (probabilistic, risky): such and such strategy choices by A and B only determine a probability distribution for the outcome.

Player A, at his first move, finds himself with either Hi or Lo, according to the whim of chance. The choices open to him are either to fold (F) or to raise (R). So this stage of a strategy for A consists of a *pair* of *conditional* moves – e.g. 'if Lo fold, if Hi raise'. In this short, rudimentary game, there are in fact no further moves to be made by A – so here this single conditional pair is one entire strategy. There are, evidently, four such conditional-pair strategies for A. B's strategies are even simpler: fold (F) or call (C): 'fold' means that B gives the 'instruction to his agent': 'If A raises, fold.' Similarly for 'call'.

The matrix below shows the expected money payoff to A if he and B choose such-and-such strategies. The rows correspond to the strategies of A. F, F denotes the strategy 'F if Lo, F if Hi': similarly for the other rows, the Lo-contingent move always being given first. The columns correspond to B's strategies of fold and call respectively.

		B	
		F	C
A	F, F	$-a$	a
	F, R	0	$(b-a)/2$
	R, F	0	$-(a+b)/2$
	R, R	a	0

Let us check one entry, in particular confirming that it is an *expected value* rather than an amount that will come about for sure. Suppose A plays F, R and B plays F. With probability 1/2 A draws Lo and so folds: in these circumstances he loses a. With probability also 1/2 he draws Hi and raises; since B folds in this contingency, having adopted strategy F, A wins a. Thus A's expected money gain is $\frac{1}{2}(-a)+\frac{1}{2}a=0$.

In every eventuality A's realised (actual) money gain is B's money loss, therefore A's expected gain is B's expected loss, and there is no need to print out B's expected money outcomes.

Later we are going to give the solution of this game. What the theory of games holds to be rational play by A requires him to use the bluff strategy R, R on some hands; or, if the game is played once only, to use this strategy with a certain *probability*.

3.3 PAYOFFS

According to element **(3)**, once each player has chosen a strategy, a probability distribution for the outcome of the game is determined. Say there are n players A, B, . . ., N, and denote their chosen strategies by α, β, . . ., ν. Then the ordered list or *vector* of these strategies, written $(\alpha, \beta, \ldots, \nu)$ determines a (probabilistic) outcome of the game. In the two-person game of Rustic Poker, for example, the two-place vector, or pair ((F, R), F) gave the probabilistic outcome: with probability 1/2, $-a$; with probability 1/2, a.

In determining a probability distribution for the outcome the vector of strategy choices determines, too, a vector of expected utilities, one for each player. Now the players of a game are assumed to have VNM preferences (element **(2)** above); hence they 'have' utility indices whose *expected values* are their maximands. What the literature calls the *payoffs* of a strategy are these *expected utilities* to the players. If the word payoff is used to refer to anything but an expected utility, it must be qualified, as when we talked about 'money payoffs' in the last section.

In the rest of this chapter and in the two following chapters we shall confine our attention to games with just two players. Let us continue to call them A and B.

Suppose that, in some game, for every pair of possible outcomes, whether risky or sure, if A prefers the first to the second or is indifferent, then B prefers the second to the first or is indifferent, and vice versa. In symbols, for all L, L'

$$L \succsim_{\mathrm{A}} L' \text{ if and only if } L' \succsim_{\mathrm{B}} L. \tag{3.1}$$

A's and B's preferences are diametrically opposed. A and B are perfect antagonists. The game they are playing is then called *strictly competitive*. Of course A and B may well feel sympathetically about issues outside the game: indeed people who play competitive games with each other often have much in common.

If (3.1) holds, we can choose VNM utility indices u_{A}, u_{B} for A, B which succeed in representing their preferences among all the game's possible outcomes and which are such that $u_{\mathrm{B}} - u_{\mathrm{A}}$ for every sure prospect. [This is quite easy to see intuitively. Can you prove it?[1]] If we do this, then, whatever the game's outcome, the sum of A's and B's utilities equals zero. Thus, *strictly competitive* games are *zero-sum* in the two player's utilities. The rest of this chapter is about zero-sum (ZS) games. These games have the most cut-and-dried properties, and their mathematical analysis is the most complete. But they are only a very special subclass of all games (see the Chart on p. ii).

A game is played for *utility*. (Better, since chance may take a hand, for expected utility.) In games whose outcomes are monetary and risky there is no presumption that a player will prefer one outcome of the game to another just because the first has a higher expected *money* value for him – that presumption reckons without risk aversion. But if each player's utility were *linear* in money – if there were no risk aversion – then that presumption would be correct. In this case it is also true that the game's being zero-money-sum (like Rustic Poker) implies that it is also zero-sum.

3.4 UNCERTAINTY AND SECURITY [18]

Nothing in the definition of a game says that when you decide on a strategy you blurt it out. If you are playing a ZS game you would be a fool to. [Why is this? Look at the Rustic Poker matrix of section 3.2 – interpreting it as a payoff matrix, a matrix of expected utilities, by supposing neither player to be risk-averse.]

Your opponent, therefore – and likewise you – will perforce choose strategies in ignorance of each other's choice. The two players are obliged by the dictates of rational self-interest to keep silent, and consequently they have to make their decisions independently. There is no possibility that they will put their heads together and choose a *pairing* of their strategies – for in their 'strictly competitive' circumstances there can be no advantage to either in doing so: if you knew my intention this could only damage me and it could never benefit me.

Zero-sum games are thus non-cooperative games in the sense that no cooperation – joint decisions on strategies – will ever *take place*. Later, we shall define non-cooperative games as those in which such joint decisions are *precluded* either by the rules or by physical

[1] A proof of the converse proposition is given in section 4.1.

constraints. Here there is no need for barriers to communication, because neither player 'wants to know': cooperation is futile, hence irrational. For definitional consistency, however, we note that a ZS game is formally non-cooperative only if cooperation is impossible.

Denote A's strategies by α_i ($i = 1, 2, \ldots$) and B's by β_j ($j = 1, 2, \ldots$). B is not a fool, so A does not know which β_j it is that B has chosen. Hence A's optimum strategy is not determined – at least an optimum in the usual sense is not. This is the consequence of *interdependence*, of the fact that A's optimal strategy depends on β_j, together with A's ignorance of B's choice. It is the fundamental problem to which the theory of two-person ZS games is addressed.

How then can A rationally go about choosing a strategy? We are confronted here with a type of choice situation radically different from those of familiar economic theory – barring the 'anomalies' of duopoly and certain externality cases with which traditional economics skirmishes briefly. We cannot just say 'A maximises his utility', or for that matter his expected utility, because he does not have a well-defined maximisation problem unless and until he knows what B will do. We need to think radically, far beyond the simple idea of maximisation. We have to ask the deep question: What can count as rationality in these unfamiliar conditions? Before we look at VNM's suggestion we remark that it would be nice if we could find a principle for choosing strategies with the following property:

> If both use the choice-principle, neither will afterwards regret having used it. (3.2)

Or, what is equivalent to property 3.2: if both used it, neither would wish to change his chosen strategy even if, having made his choice, he were informed of the other's. To erect such a principle and to expect there to exist choices by A and B that satisfy it, is to set our sights high. The essence of our difficulties is exactly that my choice varies according to yours. Yet here we are requiring choices by the pair of us that would *not* be amended if the other's were announced. In short, the principle would pick out a pair of choices that are in *equilibrium* with each other – if only such a pair exists.

This notion of equilibrium, sometimes distinguished by the term *non-cooperative equilibrium*, goes deeper than the ordinary notion of equilibrium in economics – which is a special case of it. Non-cooperative equilibrium is a property of a pair (more generally, a vector) of independently made choices or actions: a condition in which no individual has motive to change provided that (i) no other individual changes; as well as that (ii) no change occurs in the environment. (i) says that equilibrium strategies are 'good against each other'. In orthodox economics, condition (i) drops out, leaving only (ii) – either because we consider only one conscious agent, so that *all* factors relevant to his decision are environmental; or because the choices of other conscious agents, though there are some, are irrelevant, because there is no interdependence; or finally, because all the relevant choices of the other agents are known to the first.

There is a further property, deeper than property 3.2, which we might want the choice-principle to have. Consider A not merely reacting to B's choice, to the β_j he chooses, but reacting to the choice-principle that B is employing. We are led to ask that the choice-principle should have the property:

Neither should be deterred from using the choice-principle by assuming that the other is using it. (3.3)

Notice that property 3.3 refers to situations in which A has, *ex ante*, the necessary information to apply the principle to B's problem.

We now give another example of a ZS game; next we use this example to introduce the choice-principle suggested by VNM; finally, we shall check whether the VNM principle has properties 3.2 and 3.3 if it is applied to the example.

EXAMPLE: THE BATTLE OF THE BISMARK SEA

In early May 1942, U.S. intelligence believed that the Japanese were about to move a troop and supply convey from the port of Rabaul, on the eastern tip of New Britain, to Lae, to the west, on New Guinea. It might either travel to the north side of New Britain, where visibility would be poor, or to the south, where visibility would be clear. In either case the trip would take three days. The American general, Kenney, had to decide whether to concentrate his reconnaissance aircraft on the north or south side of the island. Once sighted, the convoy could be bombed until its arrival at Lae. The entries below show the expected number of days' bombing. We assume that the utilities of the two players, the U.S. and Japanese forces, are both linear in days of bombing – with positive and negative coefficients, respectively. So without loss of generality [why?] they may be taken to be simply $\pm$ the number of days, and the game is a ZS one with payoffs or expected utilities as shown.

		Japanese N	Japanese S
Kenney	N	2	2
	S	1	3

What should Kenney do? Call Kenney A and the Japanese commander B. If Kenney plays α_N (in an obvious notation), he gets 2 whatever the Japanese do. If he plays α_S his payoff varies with his antagonist's strategy – as a strategy's payoffs normally do: but he gets *at least* 1 whatever the Japanese strategy is. (This magnitude 1 is a parameter of α_S which is independent of β_j: this will be significant.) Each strategy of A has this unambiguously defined measure – the payoff *assured*, that is the largest number that the player's payoff will for sure[1] either equal or exceed: this is called the *security level* of the strategy. It is 2 for α_N and 1 for α_S. For the Japanese strategies [check this] the security levels are: -2 for β_N and -3 for β_S.

The fanciable property 3.2 of a choice-principle and the still more fanciable one, 3.3, may turn out to be luxuries that cannot be afforded: we do not yet know if there *are* any strategies, one for each, which satisfy principles having these properties; in short, we do not yet know that any equilibrium pair of strategies exists.

On the other hand, an indispensable property is that A can *use* the choice-principle without knowing β_j – because he does not. A principle which at least does this is: A should maximise his *security level.* Since the security level of each of A's strategies is independent of β_j, this *is* a well-defined maximisation problem. The principle would lead Kenney to reconnoitre north; and the Japanese to sail north. [Check this.]

The principle is pessimistic in the extreme: in assessing each strategy it considers only the worst counter (generally a *different* worst for each. There is no warrant for this as a hypothesis about what is true – especially if B has made up his mind!) It is called the *maximin* criterion for selecting a strategy – it picks the strategy whose minimum payoff is the maximum of those for all strategies:

[1] Of course, if the outcomes are stochastic, probabilistic, all we can say is that the *expected* number of days is bound to be at least so-and-so (i.e. would be at least so-and-so for every strategy of the opponent). The minimum length of bombing that might *eventualise*, the minimum amount of realised, not expected, utility, can well be lower – if for example B plays a good counter-strategy and the vagaries of the weather are against A too.

it maximises the minimum. This is the choice-principle that VNM proposed. In military terms it focuses on the enemy's capabilities rather than his intentions: the former are known, the latter can at best be guessed. This was the ruling doctrine in American tactics, and Kenney played α_N. The Japanese played β_N, were sighted after about one day, and suffered heavy losses.

3.5 SADDLE POINTS

It is no part of our project to justify the maximin principle absolutely. (By what standards could one assert that it is more rational in an ultimate sense than some rival? We shan't go into this fundamental question of decision theory.) In section 3.9, on 'mixed strategies', we draw attention to one disquieting property of the maximin rule. Meanwhile, we are going to show that the rule satisfies certain possible *a priori* requirements of rational decision in a non-cooperative situation, and leave it at that.

We have already seen that the rule is *operational.* Now we investigate whether it has property 3.2 of producing equilibrium, and the 'meta'-property, 3.3.

Consider property 3.2. In the Bismark Sea game, are the maximin strategies good against each other? The Japanese maximin strategy is β_N: α_N, Kenney's maximin, *is* best against this. Similarly, β_N is (equal) best against α_N. So α_N, β_N *are* in equilibrium: if both play thus there will be no regrets. If each were informed of the other's plans, neither would change his plans. Property 3.2 is satisfied.

Now for property 3.3. Suppose Kenney assumed the Japanese commander was using the maximin criterion. Since he knows the Japanese payoffs (by element **(4)**) he could then *deduce* that he would play β_N. He would then himself play α_N, that is, as maximin dictates. Similarly vice versa. So 3.3 is satisfied.

The next step is to see whether what is true in the Bismark Sea game – that maximin strategies are in equilibrium – is true generally. First, we need a general notation for the payoff or expected utility to A if A plays the arbitrary strategy α_i and B the arbitrary strategy β_j. It would be natural to write this as $Eu_A(\alpha_i, \beta_j)$ – but cumbersome. We shall retain all the essential information if we simply write u_{ij}. For one thing, if we want to know B's payoff it is enough to know that A's is u_{ij}, for B's is then immediately given as $-u_{ij}$ (the game being ZS). As for dropping the expectation symbol E, nothing is really lost: if the payoff happens to be sure, not probabilistic, it still has, technically, an expected value and this *is* just its sure value; thus, we may legitimately use the same symbol for expected utility and utility *tout court*.

We can now express concisely the equilibrium property of the strategy pair α_N, β_N, the property that each is 'good against' the other. α_N being good against β_N *means* $u_{NN} = \max_i u_{iN}$; β_N being good against α_N means $u_{NN} = \min_j u_{Nj}$. That is, if α_N and β_N are in equilibrium, u_{NN} is the *maximum of its column and minimum of its row*. In a three-dimensional diagram with the αs ranged along one horizontal axis, the βs along the other, and u on the vertical, there is a *saddle point* where $\alpha_i = \alpha_N$, $\beta_j = \beta_N$. u passes through a maximum as one moves through this point in the α direction, through a minimum as one moves through it in the β direction.

To recap: an *equilibrium* is a pair of strategies such that property 3.2 is satisfied. Algebraically, there is one whenever there is an entry in the payoff matrix which is the maximum of its column and the minimum of its row. Geometrically, there is one whenever there is a 'saddle point' in the three-dimensional (α, β, u) diagram.

Is there always equilibrium between the two players' maximin strategies? Does the VNM principle give in all games a pair of choices for which (3.2) is satisfied? We can see now that it certainly does not. For it is obvious that one can construct a payoff matrix (invent a game) which does not have *any* element which is both maximum in its column and minimum in its row: *a fortiori* the element given by the two maximin strategies is not. Consider the payoff matrix

$$\begin{bmatrix} 40 & 20 \\ 10 & 30 \end{bmatrix}.$$

[Check the maximin strategies and that they are in disequilibrium.] We shall return soon (section 3.8) to consider whether maximin can be saved in the face of this disturbing finding.

3.6 WAGE DETERMINATION I

A trade union and the management of a plant are submitting a pay dispute to arbitration. The board is to consist of one representative of each side, and an impartial chairman. The outcome of the hearings will be a pay increase measured, say, in pence per hour. It is assumed, with an unrealism which we shall repair later (section 5.7), that the labour and management sides' utilities are (positively and negatively) linear in the award.

The strategies of the two sides are the choice of one or another representative. Each has four candidates, whom we shall designate somewhat crudely, according to their degree of commitment to or vigour in putting their client's case, as u (unbiased), b (biased), s ('soft sell') and o (outspoken). We shall label the two sides' strategies, in an obvious notation, as $\alpha_u, \alpha_b, \ldots, \beta_u, \beta_b, \ldots$.

Experience suggests the following expected values for the award if this or that pair of spokesmen turn out to be on the board. These expectations are held by both sides.

		Management			
		β_u	β_b	β_s	β_o
	α_u	15	5	10	10
	α_b	45	40	20	25
Union	α_s	20	10	30	30
	α_o	15	20	15	15

The figures in this example are only loosely based on real cases: they are essentially illustrative, and we shall not labour the interpretation.

The payoff matrix illustrates the important principle of *dominance*. Observe that, for A (the union), α_b and α_s are better than α_u *whatever* B plays; similarly, α_b is unconditionally better than α_o. α_u and α_o are said to be *dominated* (α_u by α_b and α_s, and α_o by α_b). If one action is preferable to another in every possible 'state of the world', it is a clear principle of rational choice – known as the *sure-thing principle* – that the first should be chosen. Hence A may drop the unbiased and the outspoken candidate from consideration. B, being rational and attributing rationality to A, knows that A must discard α_u and α_o, and B can therefore confine himself to the reduced game with the payoff matrix enclosed in the broken line. But in this smaller game β_u and β_o are dominated. B therefore discards them. We are finally left with the following 2×2 matrix, which we recognise as that of the last section, and which we now border with the security levels of the two players.

	β_b	β_s	
α_b	40	20	20
α_s	10	30	10
	−40	−30	

3.7 MIXED STRATEGIES

Our argument now requires the following theorem. The theorem says that, if A's maximin payoff and B's maximin payoff are equal and opposite, then the game has a saddle point or equilibrium. (It does not say that the two maximin strategies themselves constitute this equilibrium – although that does turn out to be the case.) Now B's maximin payoff or expected utility is the same thing as his minimax expected disutility – where his 'disutility' simply means the negative of his utility. [Check this, being careful to maximise and minimise with respect to the right player's strategies.] Because the game is ZS, the latter, in turn, is B's minimax of A's expected utility. So the theorem says, equivalently, that if A maximining and B minimaxing A's expected utility u_{ji} give the same answer, then the game has an equilibrium.

**THEOREM. If $\max_i \min_j u_{ij} = \min_j \max_i u_{ij}$, then the game has a saddle point (equilibrium).

PROOF. Call the ith row minimum (security level of α_i) s_i, call the jth column maximum (B's maximum possible expected disutility playing β_j) t_j. Let i^* denote the row for which s_i is maximised (or any such row if there are more than one), i.e. $\max_i s_i = s_{i^*}$. Similarly, say $\min_j t_j = t_{j^*}$.

Call the common value of s_{i^*} and t_{j^*} hypothesised in the theorem v^*.

Consider $u_{i^*j^*}$. Suppose it is not its row minimum: i.e. $v^* < u_{i^*j^*}$. Since $v^* = t_{j^*} = j^*$th column maximum, it $\geqslant u_{i^*j^*}$.

The contradiction shows that $u_{i^*j^*}$ *is* its row minimum; by a symmetric argument it is its column maximum. That is, $(\alpha_{i^*}, \beta_{j^*})$ is a saddle point. □ **

In what circumstances does a game have an equilibrium at all? The theorem shows that it does if maximin u_{ij} = minimax u_{ij}. In the arbitration game there is, by inspection of the 2×2 payoff matrix, no saddle point. Then, according to the theorem, maximin u_{ij} must differ from minimax u_{ij}. Indeed this is so: maximin $u_{ij} = 20$, minimax $u_{ij} = 30$. In words, 20 is the most that A can be sure of, his highest security level; 30 is the least B can 'hold A down to', his lowest *hazard level* – where by the 'hazard level' of a strategy of B we mean the maximum possible expected disutility from playing it, or what he would expect to lose if A's counter-strategy were as damaging as could be.

The absence of equilibrium appears as a gap between the most A can be sure of and the least B can hold him down to. If, on the other hand, there is no discrepancy, the theorem promises a saddle point.

Suppose we could enlarge the set of strategies of each player so as to (i) raise A's best security level, (ii) reduce B's least hazard level. This would bring us nearer to equilibrium.

Now look again at the last payoff matrix. A is deterred from playing α_s by the possibility that B is playing β_b, giving the low 10 for the wage award. However, still keeping to the hypothesis that B is playing β_b, suppose that A throws a die, choosing α_b if a six comes up and α_s otherwise. Then the 40 provides a hedge against the 10. A's *expected* award is now $\frac{5}{6} \times 10 + \frac{1}{6} \times 40 = 15$. So by the device of throwing a die A has apparently found an additional strategy (of a rather new kind) in which the threat of the low payoff of 10 is mitigated. [Check that 15 is the security level of this new strategy.] A strategy like this, which determines which of the original strategies is to be played by reference to a random experiment, is called a *randomised* or *mixed* strategy. Thus, a randomised or mixed strategy is one under which each of the original, or *pure* strategies, will be played with a certain probability. (If there are more than two pure strategies some may have zero probabilities of being selected.)

A can do better still. By playing α_b, α_s with probabilities half and half he gets a security level of 25. That is, higher than any he had before. This, however, is the end of the line. If he raises the probability with which he plays α_b any further, the influence of the 20 payoff in the case of B's playing β_s begins to pull the security level down again.

All this looks a bit fishy. Nothing has changed in the world. Player A has apparently improved his position simply by a mechanised form of indecisiveness. Does the suspect character of the argument come from the introduction of expected values into it? It cannot be this, for expected values were already there. The outcomes were stochastic and the payoffs only expected utilities in the first place. (Notice, by the way, that this means that the expected award of 15 which we computed above was actually an 'expected value of expected values'. There is a two-stage random experiment: stage one chooses the strategy α_b or α_s; stage two determines the award when, with either union representative b or s confronting the management, new random factors come into play – the mood of the chairman, the current degree of government suasion, and so on.)

What is dubious is not the presence of expectations which reflect imperfect knowledge about matters outside the agent's control, but the notion that he imperfectly knows what is after all entirely up to him to decide. This kind of incertitude can only come about by a deliberate abdication of his free choice to fortune. Can this be rational? We shall return to this question in section 3.9.

Meanwhile, to recap: allowing randomised strategies raises A's security level. So it is in B's case. [Confirm this numerically.] It raises B's maximum security level, or, equivalently, decreases his lowest hazard level. Thus, it narrows the gap between maximin u_{ij} and minimax u_{ij} and brings us towards equilibrium. Indeed it eliminates the gap and establishes equilibrium in every two-person zero-sum game. This is what the following famous theorem says.

3.8 THE MINIMAX THEOREM

MINIMAX THEOREM (or 'fundamental theorem of two-person ZS games'). Every two-person ZS (finite) game has an equilibrium pair of pure or mixed strategies.

The qualification 'finite' excludes games (like pistol duels) in which the set of pure strategies (at which point to turn and fire) is continuous or otherwise infinite.

PROOF. Instead of giving a completely general proof, we take a concrete case – the arbitration game of section 3.6 in its once-reduced form in which A's two dominated strategies have been excluded. The payoff matrix of this reduced game is reproduced below. A has just two pure strategies, which makes it possible to show the steps of the proof geometrically in a two-dimensional diagram. However, the ideas of the proof are really not geometric but come from vector algebra or co-ordinate geometry. This means that the proof can easily be generalised: the force of its arguments does not depend on A's having only two pure strategies. The cost of this generality is that some elementary knowledge of vector algebra or co-ordinate geometry is needed to follow the argument. But these points will be explained as we go along. The proof, though long, is not particularly difficult.

The proof is 'constructive' – it actually determines the equilibrium pair of strategies which we claim to exist. Although this is more than is called for by the theorem, it is worth doing because it provides a technique for actually solving games. Indeed games in which, as in our illustration, one player has only two pure strategies, can be solved by going quite mechanically through the same steps we take below.

Consider, then, the payoffs in the arbitration game without A's dominated strategies:

		B			
		β_u	β_b	β_s	β_o
A	α_b	45	40	20	25
	α_s	20	10	30	30

The axes in Figure 3.1 measure the payoffs to A when he plays α_b, α_s, respectively: we denote these two variables by u_b, u_s. Each circled point represents a pure strategy β_j of B. For example, the point labelled β_o shows that if B plays β_o then A gets payoff $u_b = 25$ if he plays α_b and $u_s = 30$ if he plays α_s. Thus, the coordinates of the points correspond to the columns of the above payoff matrix.

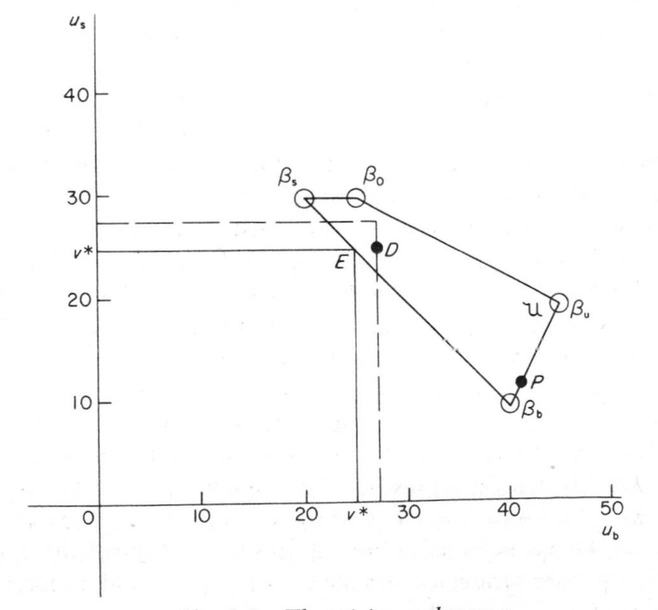

Fig. 3.1 *The minimax theorem*

The polyhedron (straight-sided closed figure) $\mathscr{U}$ whose vertices are the β_j values, is called the *convex hull* of the points β_j. It is obtained by drawing a string tightly round pins located at the β_js. (One must of course be careful not to dislodge the pins.) In another game some β_js could well be situated inside the convex hull of the β_js – it just so happens that none is in the present example.

This convex hull encloses the set of all points (u_b, u_s) got by 'mixing', i.e. taking *weighted averages* or 'convex combinations' of the points β_j. It is an elementary result of co-ordinate geometry or vector algebra that if λ is a number between 0 and 1, then the point which is situated the fraction λ of the way along the straight line from β_j to $\beta_{j'}$ has co-ordinates $(1-\lambda)\beta_j+\lambda\beta_{j'}$. That is to say, its horizontal co-ordinate is $(1-\lambda)$ times that of β_j plus λ times that of $\beta_{j'}$; similarly for its vertical co-ordinate. For instance, the point P, 1/5 of the way from β_b to β_u, has co-ordinates $(\frac{4}{5}40+\frac{1}{5}10)$ and $(\frac{4}{5}45+\frac{1}{5}20)$, or 34 and 40. [Check this by measurement.] This explains why a point like P is called a 'weighted average' of β_j and $\beta_{j'}$, with 'weights' $1-\lambda$ and λ.

The set of all weighted averages of pairs of β_js consists of boundaries of $\mathscr{U}$ like $\beta_o\beta_u$ and diagonals like $\beta_o\beta_b$. In the same way, the point whose co-ordinates are given by $\lambda\beta_j+\mu\beta_{j'}+\nu\beta_{j''}$, where λ, μ, ν are non-negative numbers adding to one, is a 'weighted average' of the points β_j, $\beta_{j'}$ and $\beta_{j''}$. Such a point will lie on the join of a vertex of $\mathscr{U}$ with a point on a boundary or a diagonal not containing the vertex. In this way the whole of the interior of $\mathscr{U}$ may be filled in. In sum, $\mathscr{U}$ consists of points whose co-ordinates are weighted averages of the co-ordinates of the β_js.

Now what such mixtures or weighted averages of the pure-strategy points β_j show is precisely the effect on the payoffs to A of B's mixing his pure strategies with probabilities equal to the weights. For example, the co-ordinates of the point 1/5 of the way from β_b to β_u are the payoffs u_b and u_s, respectively, when B plays β_b with probability 4/5 and β_u with probability 1/5 – or, in the notation we now adopt, when B plays the mixed strategy $(\frac{4}{5}\beta_b, \frac{1}{5}\beta_u)$. We check this. The point 1/5 of the way from $\beta_b = (40, 10)$ to $\beta_u = (45, 20)$ is (41, 12). On the other hand if A plays α_b he gets expected utility $u_b = \frac{4}{5}u_{bb}+\frac{1}{5}u_{bu} = \frac{4}{5}40+\frac{1}{5}45 = 41$. Similarly [check this] if he plays α_s he has expected utility 12. In short, $\mathscr{U}$ shows the payoffs to A's two pure strategies for all possible pure and randomised strategies of B.

Look at the game from B's standpoint. We shall locate his strategy (mixed or pure) which *minimaxes* A's payoff, or maximins his own. He can hold A down to at most v, that is, he has a hazard level of v (A's highest payoff is v), if $\mathcal{U}$ contains a point with both co-ordinates $\leqslant v$ and one of them $= v$. The broken line in the diagram has both co-ordinates $\leqslant 27{\cdot}5$ with one $= 27{\cdot}5$. Since it has points, such as D, in $\mathcal{U}$, by mixing his strategies to bring (u_b, u_s) to D, player B risks losing no more than 27·5. B wants to minimise this maximum loss, i.e. to find a line like the dashed one, right-angled and symmetric round the 45° line, whose v is minimal and which yet has a point in common with U. In our case this is clearly the line with corner at E.

Call the minimised v-value v^*. It is clear that, in every example, v^* will be achieved somewhere on the south-west frontier of $\mathcal{U}$ (which may, in other examples, be kinked). Generally, the point of contact will come on a facet of $\mathcal{U}$, as E does in the present case, rather than at a vertex: thus, in general it will correspond to a mixture of two pure strategies β_j of B.

We have found B's minimax (pure or mixed) strategy. Call it β^*. It gives A the expected utility or payoff v^* both if he plays u_b and if he plays u_s: thus it gives A the point E in Figure 3.1. It mixes β_b and β_s in the probability proportions $\beta_s E : E\beta_b$. Notice that the strategies β_o and β_u could not be involved because they are not on the south-west frontier of $\mathcal{U}$. This in turn is because they are dominated.

Now consider the line which 'separates' $\mathcal{U}$ from the box Ov^*Ev^*. In our case this 'separating' line is just the south-west boundary of $\mathcal{U}$. (In other cases it might be a facet of the south-west boundary, or, freakishly, a tangent at a vertex on the south-west boundary.) As this separating line or south-west boundary is a straight line in the (u_b, u_s) plane it has equation

$$x_b u_b + x_s u_s = k,$$

and with no loss of generality we can take the sum of the coefficients x_b, x_s to be unity. Substituting (v^*, v^*), which lies on the separating line, in this equation we get $k = v^*$, so the line is

$$x_b u_b + x_s u_s = v^*. \tag{3.4}$$

Now consider A's mixed strategy $(x_b \alpha_b, x_s \alpha_s)$: call it α^*. This *is* a mixed strategy because x_b and x_s add up to one, and they must be positive because the boundary slopes downwards; so x_b and x_s are probabilities.

We now show that α^* and β^* are equilibrium strategies, that is, good against each other. First, α^* against β^*. Both the co-ordinates of β^* are v^*, i.e. A gets v^* whichever pure strategy he plays against β^*; *a fortiori* he does so whatever mixture of them he plays. α^* is one such mixture, so he can achieve no more against β^* than he does by α^*.

Also, β^* is good against α^*. Since our line (equation 3.4) separates $\mathscr{U}$, that is to say $\mathscr{U}$ lies all on the far side of it, we have $x_b u_b + x_s u_s \geqslant v^*$ for all points (u_b, u_s) in $\mathscr{U}$. But this means that A's expected utility from $(x_b\alpha_b, x_s\alpha_s)$ is v^* or more whatever point in $\mathscr{U}$ player B picks, i.e. whatever strategy B plays. On the other hand for the point *E*, i.e. for B's mixed strategy β^*, it is exactly v^*. Hence β^* is good against α^*.

We have shown that there is an equilibrium pair of strategies, (α^*, β^*). □

We have also almost shown something more. A by-product of our proof was that β^* is B's minimax. Our last result is that α^*, its equilibrium counterpart, is A's maximin.

If A plays α^* he cannot get less than v^*, i.e. v^* is the security level of α^*. Suppose that some other strategy α' (pure or mixed) has a higher security level $s' > v^*$. Then if A plays α' and *B* plays β^*, A gets at least s' and therefore more than v^*. But if A plays α^* and B plays β^* we have seen that A gets v^*. So then α' is better than α^* against β^*, which is a contradiction. Therefore α^* is A's maximin strategy.

The equilibrium strategies whose existence we have established are maximin and minimax. Hence the name of the theorem. [Show that it is a corollary of what we have proved that A's and B's maximin payoffs must be equal and opposite.]

Finally we draw attention to a curious fact about the solution. If B plays his equilibrium strategy β^*, each of the individual pure strategies involved in the optimum counter-strategy α^* yields the same payoff. Symmetrically, if A plays α^* then the pure strategies which enter the optimal counter-strategy β^* also severally yield the same payoff. (This result does not depend on how many pure strategies appear in the equilibrium mixed strategies: this number may be greater than two in other games.)

[Show by the geometrical construction of the above proof, and measurement, that the solution of the arbitration game is: $(\frac{1}{2}\alpha_b, \frac{1}{2}\alpha_s)$, $(\frac{1}{4}\beta_b, \frac{3}{4}\beta_s)$.]

3.9 INTERPRETATIONS OF MIXED STRATEGIES

REPEATED GAMES

Suppose that the same arbitration game between union and management took place every year; and suppose that the union chose a pure strategy once and for all and used it every year. The management would learn it sooner or later – though perhaps too rapid an inductive inference would not be rational in the circumstances, as the management should be on its guard against a trick. If the pure strategy were α_b, management would (before long) counter with β_s and thereafter hold the union down to 20. If α_s, it could hold the union down to 10. Clearly, the union would soon wish to abandon its commitment to a constant pure strategy.

On the other hand, suppose that instead of adopting a single pure strategy, over a number of years the union played α_b one year out of two and α_s in the other years. Suppose, furthermore, that each year the union decided whether it was to be an α_b year or an α_s year by tossing a coin. The management could then never know which style of representative it was to face – only that the probabilities were half and half. It would have no means of correlating its counter-strategies with the union's strategies. (If on the other hand the union played each strategy not randomly but alternately, it would gain little – for though the management might at first be confused it would eventually cotton on.)

This logic of randomisation is familiar from games of bluff. Take Rustic Poker. If I always played (F, R) – fold if Lo, raise if Hi – my opponent, once he realised this, would always know my hand and by playing F he could hold me down to 0 in the long run. Similarly, if I always bluffed – (R, R) – by countering with C he could again hold me down to 0.

What is A's optimal strategy in this simple poker according to the theory of games? Consider again the payoff matrix on page 39. We first observe that (F, F) and (R, F) are dominated: it is always better to raise on Hi. This leaves rows 2 and 4. The game may be solved, in particular numerical cases and in general, by our method of proof of the Minimax Theorem. The general solution is that A plays $(p(\mathrm{F, B}), (1-p)(\mathrm{B, B}))$, with

$$p = \frac{2a}{a+b}, \quad \text{and then} \quad v^* = \frac{b-a}{b+a}\, a.$$

That is, the bigger is the bet relative to the ante, the more A should bluff. For an ante of 1, if $b = 3$, the bluff probability $1-p = 1/2$; if $b = 9$, $1-p = 0{\cdot}8$. The theory assumes that A knows how to keep a straight face.

ONE-OFF GAMES

If a game is played just once we cannot interpret the probabilities in a mixed strategy in the way we have just done, as long-run relative frequencies. Hence we cannot rationalise the use of a mixed strategy as an attempt to frustrate an intelligent opponent's learning your intentions by experience of playing with you.

A player using a mixed strategy performs a simple random experiment – throwing a die, spinning a wheel – and plays one or another pure strategy as its outcome dictates. That is all. So there comes a point when, after all, a *pure* strategy is selected. But then why not choose the best pure strategy in the first place? The physical experiment has in no way affected the prospects of the pure strategies.

The experiment appears not merely superfluous: there is a chance that the wheel will indicate a pure strategy definitely worse than some other. Resort to a random device to do one's deciding for one seems positively irresponsible. Imagine a decision to use a nuclear weapon made by a roulette wheel. Yet this is what the theory may recommend when it recommends the maximin strategy, be it pure or mixed.

The precept cannot be dismissed out of hand. Unquestionably maximum security levels can be raised, and the defence we have offered of focusing on security levels – their equilibrium property – is a substantial one. Can the paradox be resolved? The issue is deep and complex, but the following argument may go some way towards clarifying it.

It has already been said (section 3.2) that using *strategies* rather than making a sequence of 'point' decisions as you go along does not penalise or hamper you (provided your brain is a perfect machine). For a strategy anticipates (hypothetically) every possible situation in which you may find yourself. So, as your agent follows through an optimal strategy you can never regret having handed it to him. At each stage of the game you would wish to do, if you were there in person, what you have instructed him to do in those circumstances.

But if you have used the maximin expected utility principle and have been led by it to choose a mixed strategy, you *can* regret it.

Looking at the arbitration game and its optimal mixed strategy $\alpha^* = (\frac{1}{2}\alpha_{\text{l}}, \frac{1}{2}\alpha_{\text{s}})$, there is a chance of 1/2 that A will be committed to playing α_{s} – with a security level of 10 as against α_{l}'s 20. We have been led into a contradiction.

Prima facie, the contradiction may be due to either of two things: (i) the artificial tacking on of the random experiment at the beginning of the game. Our argument to 'no regret' refers to a given game with given rules and a definite sequence of moves. We have, however, mucked about with this game in reaching the contradiction. (ii) the principle for choosing a 'best' strategy. Perhaps a maximin strategy is not 'optimal' in the sense required by the argument to 'no regret'.

It seems possible to rule out (i), for exactly the same trouble arises even if the original game happens to have an initial coin-toss or some such thing written into its rules. Such a coin-toss would be one of the game's 'chance moves' like any other (except that it does not affect subsequent possibilities for the players). We may conclude that the contradiction casts doubt on the validity of the maximin principle as an optimisation principle.

Apropos, we note that no trouble of this kind arises if the players use the *Bayesian* principle for making decisions with imperfect knowledge. According to this principle they assign probabilities to each other's playing one strategy or another, then maximise the simple expected value of their utilities (the expectation being with respect to the other's choice of strategy as well as with respect to objective chance elements in the game.) Here there is never advantage in using a mixed strategy. However, the Bayesian principle patronises the player's opponent, treating him as to a degree predictable rather than as a fully rational being actively engaged in out-manouevring the first player. Furthermore, since the players' subjective probabilities for each other's strategies may be anything whatsoever, there is nothing that conduces to equilibrium. Here we leave this discussion.

4 Two-person Nonzero-sum Non-cooperative Games

4.1 NONZERO-SUM GAMES

The games we consider in this chapter differ from two-person zero-sum games in one respect only. They are not 'strictly competitive'; the players' utilities do not sum to zero. They are still for two players, and there is still no cooperation. That is, the players make their decisions independently, they do not choose a pair of strategies together.

Although the games of this chapter are like zero-sum (ZS) games in that in both types of games strategy choices are made independently, the interpretation of this lack of collaboration is different in the two cases. Zero-sum games are spontaneously non-cooperative, for there is no possible motive for collaboration. Here, however, we shall see there is. For this reason, if a nonzero-sum (NZS) game is played non-cooperatively it is so played because cooperation is either impermissible or impossible.

With the relaxation of the ZS assumption we greatly enlarge the reference of the model. This allows the concepts and methods of game theory to be brought to bear on certain classical economic problems which have for a long time proved intractable to traditional methods.

Zero-sum games are rare outside the parlour or casino. For a game to be ZS the players' preferences have to be 100 per cent, diametrically opposed. It is not enough that they should have reverse preferences among the sure outcomes of the game. Their preferences must also be reverse for lotteries over the game's outcomes. This point needs some explanation.

For a game to be ZS means that we can find valid VNM utility functions for the two players, u_A and u_B, such that for any lottery L over the game's outcomes, $u_A(L)+u_B(L)=0$. (Remember that we are using the letter u for an *expected* utility as well as for the utility of a sure prospect.) Now suppose L' is some lottery which A disprefers to L. Then $u_A(L) > u_B(L')$. But as the game is ZS, $u_A(L') + u_B(L') = 0$. Hence $u_B(L) < u_B(L')$, that is, B prefers L' to L.

It follows that a game is NZS as soon as there is one pair of lotteries L, L', such that $L \succ_A L'$ and $L \succsim_B L'$ – that is, A prefers L to L' and B would as soon have L as L'. Equivalently, it is NZS if for the pair of players some L is 'Pareto-better' than some L'. Herein lies the motive for the cooperation which is barred in the games of this chapter.

It is worth clinching all this with an example in which, although preferences are opposed for sure prospects, they are not diametrically opposed, that is they are not opposed for all lotteries. A game which has these prospects as outcomes is therefore NZS.

Let x, y, z be the sure prospects, and let u_A, u_B be some valid VNM utility indices for A, B with the following values:

	x	y	z
u_A	10	15	20
u_B	20	18	10

Preferences for sure prospects are opposite. But it is easy to check that if we take, say, $L = y$, $L' = [0{\cdot}6x, 0{\cdot}4z]$, then $L \succ_A L'$ *and* $L \succ_B L'$. We have here taken one of the lotteries, L, to be a 'degenerate' lottery, that is simply a sure prospect, but even in spite of this we have found an accord of preferences. We may interpret what has happened here as follows. B only slightly prefers x to y, while he greatly prefers y to z. Consequently – or rather, this is exactly what is meant by slight and great preferences – a mix of x, z containing even quite a lot of x, is dispreferred to y. L' is such a mix, with its 0·6 probability of x. For A, for whom y is half-way up in preference from x to z, any mix – such as L' – containing more than 50 per cent of x is dispreferred to y.

The utilities in this example, u_A and u_B, do not sum to zero. For the degenerate lotteries x, y, z their sums are 30, 33, 30, respectively. This is itself does not show the game is NZS, for the latter means that there are no valid pairs of utility functions that sum to zero, and u_A and u_B are only one particular valid pair. However, the only legitimate transformations of u_A and u_B are positive linear ones. It is possible to deduce from this [do so] that if any legitimate zero-summing indices are to be found, the given indices u_A and u_B must vary linearly with each other. We conclude that if u_A and u_B are any valid utility functions, then if, as here, u_A varies nonlinearly with u_B, there exists some probability mix of outcomes which is Pareto-superior to some other.

The reason why it is reasonable to treat parlour games as ZS (provided they are zero *money* sum) is that since the stakes are normally small compared with the fortunes of the players, the players' utility functions are approximately linear in the money outcomes; for there is negligible risk-aversion over trifles (Observation 2, section 2·5). Hence, they are approximately linear in each other.

In a wage dispute the union's possible gains may be considerable relative to its members' wealth. Thus over the range of possible awards the union's VNM utility function may be significantly curved. If, as seems plausible, the management's is more nearly linear, the game is NZS – assuming, that is, that the dispute is over shares of a fixed money sum. In any case, once one admits risk-aversion – curvature in the utility functions – it would be a sheer fluke if one curved utility function were a linear transformation of the other.

There is a quite different reason for not treating wage settlements as zero-sum between workers and employers. However competitive are the interests of the two sides *given* the sum to be distributed between them, they may coincide in wanting this joint return to be as large as possible. We shall see later (sections 4.4 – 4.8) that there are many economic situations in which the common interest of the players is, as it may be here, much more substantial than some slight nonlinearity of one's utility gain in terms of the other's utility loss when a fixed money sum is to be shared between them.

Many things change as we go from ZS to NZS games.

(i) First and foremost, it is no longer impossible to gain from cooperation. How cooperating players set about realising this gain is the subject of the next chapter. For the rest of this one, however, we discuss only those NZS games in which cooperation is forbidden by the rules or, what comes to the same as such a prohibition issued to rule-abiding players, is physically impossible.

(ii) Secondly, it is no longer necessarily damaging to let one's strategy be known. So far from playing close to the chest, it may be advantageous to advertise one's intentions. Thus it is with 'threats'.

(iii) Equilibrium strategies are no longer necessarily maximin for the two players, and maximin strategies are no longer necessarily in equilibrium.

(iv) Certain relations between equilibrium strategies when these are not unique are different in NZS games. As we left out these properties and relations in describing ZS games, we continue to do so.

These are the changes. On the other hand, the fundamental theorem of two-person ZS games generalises. It has been proved that *every (finite) non-cooperative two-person game has an equilibrium pair of pure or mixed strategies.*

4.2 THE PRISONER'S DILEMMA

This is the most famous of all nonzero-sum games. It has economic applications of truly fundamental importance. We begin by giving two alternative numerical versions of its payoff matrix, although we shall not refer to the second for some time.

I	β_N	β_C
α_N	(9, 9)	(0, 10)
α_C	(10, 0)	(1, 1)

II	β_N	β_C
α_N	(2, 2)	(0, 3)
α_C	(3, 0)	(1, 1)

First of all note that each matrix entry is now a pair of expected utilities (A's is written first). One has to write down a pair because the game is NZS, so B's payoff is not deducible from A's.

In Tucker's original story, which gives the game its name, a District Attorney is holding two prisoners in separate cells. He is sure that they have committed a crime together, but he has not enough evidence to convict unless he can get at least one of them to confess. He has each of them brought in in turn and makes each the following proposition (which establishes the payoffs.)

(a) If neither confesses (N = 'not confess') he will book both of them on some trumped-up petty larceny or illegal possession charge, in which case each may expect to get a short sentence. Let us choose the utility origins and units so that for each the utility of the maximum sentence is 0 and that of getting off altogether is 10. This involves no loss of generality, for we can always choose the origin and scale of a VNM utility function as we please (cf. section 2.4). The essential questions raised by the game do not involve differences between the two prisoners' utility functions, so for simplicity's sake we shall take them to be the same as each other for all outcomes. Suppose then that the probability of a light sentence brings the utility of each down to 9. This gives the payoff pair (9, 9) for the strategy pair (α_N, β_N) in matrix I.

(b) If both confess (C) their cooperative attitude will be taken into account and they will (probably) get less than the maximum sentence: say this probable prospect has utility 1. This gives the payoffs (1, 1) for (α_C, β_C).

(c) Finally, if one confesses and the other refuses to talk, it will go very hard with the latter (utility 0), while the former will get off free for turning State's evidence. This establishes the bottom-left and top-right payoff pairs.

The prisoners are left to ponder their decisions in isolation. Each knows that the other has been offered the same deal as himself. Note that if this were not so, the game would not possess property (4) of section 3.1, perfect knowledge of the rules. By the same property each also knows the expected utilities that face the other. In short, each knows the payoff matrix I.

The solution of the game is mechanical and easy. It is this very obviousness and incontrovertibility of what at the same time profoundly shocks our common sense that has excited so much notice.

For each player, the C strategy *dominates* the N strategy. That is, whichever strategy the other may be playing, C is the best counter. [Check this.] By the overwhelming logic of the 'sure-thing principle' (section 3.6) each player *must* therefore choose C. *A fortiori* (C, C)[1] is also an equilibrium pair. Indeed it is quite easy to show that it is the only equilibrium pair in pure or mixed strategies. (The prisoners may, if they wish, avail themselves of fair wheels supplied by the D.A.'s office.) The important thing, however, is that (C, C) dominates.

But (C, C) is terribly unsatisfactory. The sum of utilities is very submaximal. If, properly, you are uneasy about adding utilities, observe that it is also 'Pareto-improvable' – by (N, N). Yet self-interest points unmistakably to (C, C), by dominance. Let us put the matter more graphically. Suppose A begins by hypothesising: 'B will play N, thinking that I myself will'. A next realises that if B does think this, he will on reflection change his strategy to β_C – his clear self-interest demands it. But in that case A may as well save something by confessing. He opts for C. Thus, even if both begin, with the best will in the world, by contemplating a mutually beneficial solution, they are both soon driven to a mutually disastrous one.

For a moment it seems that this argument is unconvincing in one

---

[1] We shall write the strategies as C, etc. rather than as α_C, etc. when there can be no confusion.

respect, and that the prisoners may yet escape from the catastrophe of (C, C). Once A has concluded that B is going to confess, would he not be deterred from switching from N to C by compassion for his partner? By switching, A will only gain one 'utile'. If he has a shred of humanity will he not refrain from robbing his partner of nine utiles for the sake of the one that he will gain? But this apparent escape is no good, for two reasons. First, VNM utility units are not interpersonally comparable. A may think – and it may perfectly well be true – that B feels much less strongly about the whole prospect of imprisonment than he does. Although both have the same VNM utility functions, it certainly is possible that the most, or least preferred outcome, would produce very different absolute degrees of suffering for the two. Secondly, the utilities in the matrix in any case incapsulate the reactions of the individuals to the total situation. Thus A's (slight) preference for the (C, C) outcome over the (N, C) one already takes account of the fact that B will get a heavy sentence too. His concern for his partner, if he has any, is what makes his utility 1 rather than (say) 2. (For this reason, we should, strictly, have spoken of 10 as the utility A assigns not to his own freedom but to the freedom of both of them; similarly for the other assignments.) For A to alter his choices because of the relative sizes of his and B's utility indices, when his utility index already contains a response to the relative nastiness of his and B's fates, would be to double-count.

The prisoner's dilemma is the archetypal choice problem in which, contrary to the doctrines of liberal economics, the group interest is *not* furthered by the independent pursuit of individual interests. Such situations provide a case for intervention from a higher level, from which vantage not only can all the possibilities be clearly seen (*that* is not what is lacking here), but also the most favourable ones can be co-ordinated.

Prisoners' dilemmas are ubiquitous in the economy. The over-exploitation of common property resources is a (C, C)-type solution of a prisoner's dilemma with many players: whether others graze a common pasture little or much, it benefits each individual to graze it intensively. The extinction of animal and fish species used for food can be explained in the same way. So may traffic congestion. [Can you see this?] All these cases have a strong whiff of externalities about them: more of this later in the chapter. In all these cases, too, there is a price missing or one that is too low. This prompts the conjecture that prisoners' dilemmas in the economy could be dealt with by modifying the price mechanism. We shall return to this question too.

Cost inflation produced by the independent submission of claims by several competing unions may also be analysed as a (C, C)-equilibrium. We sketch a crude agrument which is meant only to show the prisoner's dilemma in the situation. Suppose real national income and its distribution between total wages and total profits are given. Let the labour force be organised into two unions, and imagine each union deliberating whether to press a low (N) or a high (C) claim. If A does the former and B the latter there will be an (N, C) payoff pair like (0, 3) in matrix II. The preferability for both of the (2, 2) of (N, N) over the (1, 1) of (C, C) reflects the possibility that in case (C, C) the government will use a policy of deflation and unemployment to combat the resulting inflation.

Politics abounds with examples of prisoners' dilemmas: disarm (N) or arm (C)? Use your vote in a non-marginal constituency (N), or don't bother (C)?

Consider one final sceptical objection to the solution proposed in the theory. Might one not expect the prisoners to *trust* each other to play N? (More precisely, each to think: 'He is going to be nice and play N in the expectation that I am – so I think I'd better.') The question is the wrong one. Rather, in a normative theory of decisions we must ask: would it be *rational* in each to trust the other? If this may be seen as a telepathetic form of cooperation, or as obedience to a moral rule governing the behaviour of criminals remanded in custody, it may be subsumed so far as game theory is concerned under the cooperative games we are going to consider later. If there is no hint of tacit understanding or prior obligation and we ask – for the last time – whether such trusting behaviour is not also individually rational for prisoners kept in total isolation, the answer must be no. C is rational by inescapable logic once we accept (a) the utility assignments and (b) the sure-thing principle.

4.3 REPEATED DILEMMAS

If B will play β_C, A's best strategy is α_C. If B will play β_N, α's best strategy is once again α_C. If the prisoner's dilemma is played non-cooperatively, that is, under the conditions we have been considering, these two facts make α_C dominant for A. If on the other hand there should be any kind of understanding, tacit or explicit, that both players will play N, the fact that α_C is best against β_N still provides A with a motive to depart from the understanding and play α_C. In these circumstances we shall call the choice of α_C a *double-cross*.

Suppose a non-cooperative game of the form of the prisoner's dilemma is played 100 times in succession. Then the repeated use of N by both players constitutes a kind of quasi-equilibrium. Suppose that the players have, somehow or other, got into an (N, N) groove. At the end of game t, A contemplates double-crossing B at $t+1$; but he argues that this will induce β_C at $t+2$, so he would be forced to play C himself at $t+2$. In one play he would have more than wiped out his transient gain. [Check.] So he sticks to α_N.

For the 100-fold game, Luce and Raiffa [18] suggest the following strategy, denoted by $\alpha_N^{(t)}$. A plays C from t on; up to t, he plays N as long as B plays N, and as soon as B should deviate from N, player A switches to and thereafter sticks to C; t would perhaps be somewhere in the 80s or 90s. In a variant of this strategy, A retaliates in short bursts if B plays C, with the object of teaching him a lesson – to stick to N. Notice that though there is a kind of 'temporal collusion' within this 100-long sequence of games, the sequence considered as a single 'supergame' is entirely non-cooperative, as A has to make up his mind entirely on his own to use the supergame strategy $\alpha_N^{(t)}$, or its variant, or yet another.

It is possible to show that every pair of equilibrium strategies in the 100-fold dilemma considered as a single non-cooperative supergame results in the repeated play of (C, C). In view of the plausibility of $\alpha_N^{(t)}$, this casts doubt on the worth of the equilibrium notion for singling out 'solutions' of NZS games. It is worth remembering that in the one-off dilemma, since C was dominant for each player, we did not in fact need to justify (C, C) as a solution by the much weaker equilibrium property – although (C, C) does, by virtue of dominance, have this property too.

On the other hand, the justification of $\alpha_N^{(t)}$ which we gave involves an inductive learning process, each player being encouraged to continue with N by experience, which progressively reassures him that the other will do so too. (In the case of the variant, he is also *dis*couraged from straying from N.) This destroys the element of uncertainty – though not of risk (cf. section 1.1) – which was part of the *raison d'être* of the idea of non-cooperative equilibrium.

Economic games, unlike military ones, usually are repeated many times. For instance, the game of oligopoly to which we now turn.

4.4 OLIGOPOLY I [25]

We shall assume that the reader is acquainted with one or more of the standard pre-game-theoretic treatments of oligopoly (see e.g. [4]). It is perfectly possible to follow the present discussion in any case, but knowing the difficulties the classic theorists met adds spice to the story of game-theory's new attempt on the old problem. The essential point is very simple, but very difficult to deal with. An oligopoly is a market with few sellers; an oligopolistic industry, one with few firms. Like most of the standard literature, we shall confine ourselves to the special case of oligopoly, *duopoly*, in which the number of sellers is two. In a duopolistic industry, the output (or selling price) chosen by one firm – B say – has a significant effect on the market possibilities facing the other. (Compare perfect competition.) Hence, unless the two firms collude, the other firm, A, cannot make a fully informed decision aimed at, for instance, maximising his profit. The outcome depends on both their decisions, but each is ignorant of the other's. What to do?

We shall work with a simple numerical model. Let the average costs of the two firms A and B be given as

$$c_A = 4 - q_A + q_A^2, \qquad c_B = 5 - q_B + q_B^2, \tag{4.1}$$

and the market demand function as

$$p = 10 - 2(q_A + q_B). \tag{4.2}$$

The average cost curves are U-shaped; one firm is somewhat more efficient than the other. The demand curve slopes downward. There is no product differentiation. The firms' assets are not specified: we assume these assets to remain unaltered as the outputs vary, i.e. the analysis is Marshallian 'short-run'.

The strategies of the two duopolists consist of independently choosing levels of *output*, which they produce according to equation 4.1 and sell as best they can – i.e. according to equation 4.2. Thus, they will play a 'quantity game'. In making outputs their decision variables they are like Cournot's duopolists, rather than Bertrand's, who choose their selling prices.

There are two minor snags in representing this duopoly situation as a game. First, unless output is indivisible, the pure strategy sets are infinite. All games we have dealt with so far have been finite, i.e. had finite sets of pure strategies (though their randomised strategy sets were infinite since the mixing probabilities could be anything.) To deal with this we simply discretise, picking out a finite number of pure strategies for each spanning the interesting range. It is clear that if the payoff surface, in a three-dimensional diagram with q_A, q_B on the horizontal axis, is smooth, we can approximate the original game arbitrarily well by using a finer and finer grid of discrete strategies. So little harm is done by discretising.

Secondly, we have not specified the utility functions of the duopolists. Profit is the traditional maximand, so to be able to compare our results with classical ones we assume that the VNM utility of each is simply his money profit. For a firm which faces risky prospects this amounts to assuming that the firm prefers more sure profit to less, and that it has no risk aversion. However, risk plays no part in the present model, so taking utility equal to profit has no implications for risk aversion, and merely implies that the firm prefers more (sure) profit to less.

Let the possible values of q_A be 0·92, 0·94, 1·17, and those of q_B be 0·41, 0·74, 0·94. (We shall see why these particular values in a minute.) It is straightforward to compute the following payoff matrix:

		q_B		
		0·41	0·74	0·94
	0·92	$(3{\cdot}14, 1{\cdot}06)^{JM}$	(2·53, 1·38)	(2·16, 1·26)
q_A	0·94	(3·16, 1·04)	$(2{\cdot}54, 1{\cdot}35)^{EQ}$	(2·17, 1·23)
	1·17	(3·21, 0·85)	(2·44, 1·01)	$(1{\cdot}84, 0{\cdot}78)^{EP}$

The reader should check that, among these strategies, (0·94, 0·74), marked EQ, is the only equilibrium pair. (So it is, incidentally, considering all pure strategies.) However (0·92, 0·41), marked JM, is the *joint maximum* pair, the pair that makes the sum of the two firms' profits a maximum (it just shades (0·94, 0·41) to further places of decimals.) The duopolists stand to gain from colluding to establish this outcome. There is a sense of having been here before. Consider on its own the top left-hand submatrix marked off by the dotted line, and check that in this subgame 0·94 and 0·74 are dominant for A, B, respectively. The top left-hand submatrix is remarkably like a prisoner's dilemma. Dominance forcibly repels the players from their joint maximum position. It is not quite one, for the true prisoner's dilemma of section 4.2 had the yet stronger feature that (N, N) was 'Pareto-better' than (C, C); here this is not so, and B only *stands* to gain from JM – he would actually gain only if there were some kickback forthcoming from A.

Notice in passing the pair marked EP for *efficient point*. Suppose that (a) each firm maximised its profit at a price he took as given; and that (b) this price were such that the resulting outputs just cleared the market. This would be the short-run equilibrium position if the two duopolists were merely two groups of firms forming the low-cost and high-cost 'halves' of a competitive industry. But they are duopolists, not collections of competitors, and there is no reason whatsoever to expect price-taking behaviour from them spontaneously. The price, 5·76, would have to be imposed by some planning ministry. The pair (1·17, 0·94) is 'efficient' in that it satisfies the rule 'marginal cost equal to price', which, if other conditions are satisfied too, establishes a level of output in our industry consistent with an optimal allocation of resources in the economy as a whole. Note that, because we have assumed the firms' assets to be fixed, this is at best a short-run criterion. If, at $p = 5{\cdot}76$, these outputs yield positive (supernormal) profit, that is a signal that capital should be reallocated from elsewhere to our industry.

Three 'solutions' to the duopoly game have been aired: EQ, JM and EP. Of these, EP can only be expected to come about by outside intervention, and JM only by monopolistic collusion. As for EQ, in the complete 3×3 game it is not dominant, and that it is a non-cooperative equilibrium pair is not only, as we have argued above, weak grounds for judging it to be rational, but weak grounds too for thinking that it will come about.

4.5 OLIGOPOLY I, CONTINUED: NON-COOPERATIVE EQUILIBRIUM AND COURNOT'S SOLUTION

In Cournot's mid-nineteenth-century theory of duopoly, in each period firm A makes its output choice on the assumption that B's output will be unaffected by A's last-period decision. In other words it flies in the face of the very essence of duopoly. For, whatever rule B has for choosing its output, its market possibilities and hence its choice must in general be affected by A's output. This interdependence means that the assumption of Cournot's firm will in general be wrong. It seems unlikely, therefore, either that (a) it will do itself much good; or that (b) it will persist in a repeatedly falsified hypothesis. But before we dismiss the theory out of hand, we shall investigate whether there is an 'equilibrium' in it. What we shall be looking for is an ordinary economic equilibrium.

The game model is static – in its one-off version. Even in a 100-fold game, the 'normal form' analysis in terms of 100-move-long non-cooperative strategies is static. (The dynamic interpretation which we gave amounts to a consideration of this supergame in 'extensive form' (cf. section 3.2).) Cournot's model, on the other hand, is an explicit dynamic one in discrete (period) time.

This treatment gives Cournot the opportunity to banish, somewhat imperiously, uncertainty. Although decisions are made simultaneously in each period, firm B's decision is subjectively given for A as a function of A's previous one. This function may be called a 'perceived reaction' function. (It is also called a 'conjectural variation' function.) The particular perceived reaction function that Cournot takes says that there is *no* reaction. Other theorists using the same general approach have postulated different, nonzero conjectural variations (cf. Sweezy's 'kinked demand' model below). If A believes that B's output will not change on account of A's last move, he might just as well believe simply that it will not change – for there is nothing else in the model that could affect it. Thus, A can determine *all* the parameters of his problem as given functions of his own decision. He has an ordinary maximisation problem. So, similarly, does B. We can therefore look for an ordinary economic equilibrium.

Suppose the duopoly game has a game-theory style, 'non-cooperative' equilibrium. Call this pair of output levels (q_A^*, q_B^*). At these output levels each player's output is, by definition of non-cooperative equilibrium, good against the other's. As we have assumed that the players' utilities are their profits, this means that each is profit-maximising by producing at the starred level given that the other is producing at the starred level. Now turn attention back to Cournot's duopolists. Each of them chooses his best, i.e. profit-maximising output on the assumption that the other is going to hold his output constant at last period's level. It follows that if Cournot's duopolists once choose to operate at the non-cooperative equilibrium levels q_A^*, q_B^*, then next period each will wish to leave his output where it is. Furthermore, the zero-reaction hypothesis on which he bases his calculations will for once be correct.

Hence (a) the game-style equilibrium is also a Cournot equilibrium, and (b) at it, the Cournot duopolists' zomby hypotheses are right, for the wrong reasons.

Nor is this all. Cournot's model, somewhat surprisingly, is *stable*, that is, the variables converge through time to their equilibrium values. Very much more by good luck than good judgement the two firms stagger their way towards (q_A^*, q_B^*). Figure 4.1 illustrates this. Call A's and B's actual (cp. perceived) reaction functions f_A, f_B. Suppose for the moment that these actual reaction functions cross the way round they do in the diagram. Begin by choosing an arbitrary output a' for A. Then B chooses b' next period, A chooses a'' the next, B chooses b'', and so on. The arrowed line tacks towards (q_A^*, q_B^*).

It is vital that the curves do not cross the other way. The main points of a proof that they do not are as follows. If B is not in the market, and so produces nothing, A will choose some positive output, $\bar{q}_A$ say. If B comes on the scene, under extremely weak conditions he too will produce *something*, say $\tilde{q}_B$. So we have shown that A's reaction curve meets the q_A axis at $(\bar{q}_A, 0)$, and B's there lies above it, at $(\bar{q}_A, \tilde{q}_B)$. Similarly at the other end. Since the reaction curves are certainly continuous they must have a stable crossing point (q_A^*, q_B^*). (We omit the conditions for them to have no more than one.)

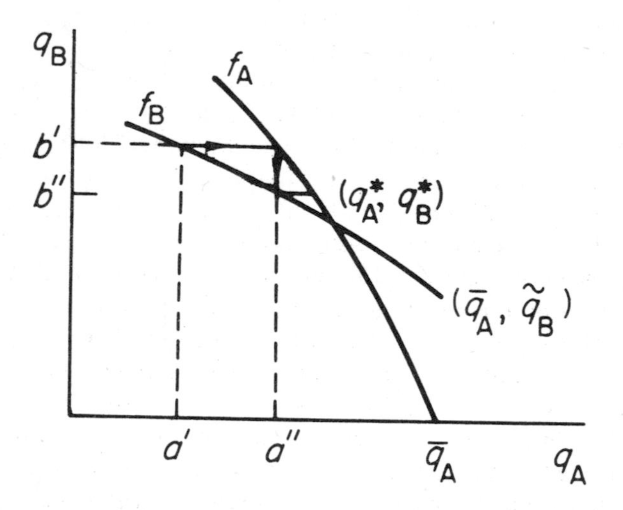

Fig. 4.1 *Cournot's duopolists*

The payoff matrix clearly must have a corresponding stability property. Given an arbitrary strategy $\alpha_{i'}$ for A, let $\beta_{j'}$ be the B strategy that maximises $u_B(\alpha_{i'}, \beta_j)$ – that is, the one that gives the maximum B-component of the payoff pairs in the $\alpha_{i'}$ row. Similarly, let $\alpha_{i''}$ maximise $u_A(\alpha_i, \beta_{j'})$ over the α_i – given the maximum A-component of the payoff pairs in the $\beta_{j'}$ column. Then the sequence of strategy-pairs (i', j'), (i'', j'), (i'', j''), etc. converges to the equilibrium pair (i^*, j^*). [Check this in the payoff matrix on p. 67.]

How can we locate the output pair which is both a game and a Cournot equilibrium if we allow q_A, q_B to take all non-negative values? For A, profit is maximised on the Cournot hypothesis that q_B is a constant if the *partial* derivative $\partial\pi_A/\partial q_A = 0$, where π_A denotes A's profit level; similarly for B. Now $\partial\pi_A/\partial q_A$ involves both q_A and q_B. So does $\partial\pi_B/\partial q_B$. We must therefore look for q_A^*, q_B^* such that both the partial derivatives vanish when evaluated *at* these q_A^*, q_B^*. If we find such q_A, q_B, we are home. [Check that the solution is $q_A^* = 0{\cdot}94$, $q_B^* = 0{\cdot}74$.]

Cournot's proposed solution to the duopoly puzzle is clearly unsatisfactory. It models the behaviour not of rational economic agents but of imbeciles. They learn nothing, clinging in spite of overwhelming counter-evidence to their expectations of zero reactions. In forming their hypotheses in the first place they fail to notice the glaring contradiction in them that would be apparent if they tried for a moment putting themselves in the other's shoes. Not only Cournot but his successors Bertrand, Edgeworth, Stackleberg and Sweezy failed to come to terms with the problem of rational duopolistic decisions: whatever is good for me is good for you; I cannot therefore attribute *mechanical* reactions to you when I myself am searching for an *optimal* rule.

Game theory does come to terms with this puzzle. It succeeds in cutting a hard old analytical knot and clearly expressing the essentials of the problem, whose symmetry of minds, the source of the difficulty, it highlights. Whether non-cooperative equilibrium is an acceptable *solution* is another matter. It does treat the participants as rational men who recognise, among other things, the rationality of the other. But our discussion of (C, C) as a 'solution' of the prisoner's dilemma both in its one-off version, and more especially in its serial version, has made the non-cooperative equilibrium criterion look somewhat questionable.

It is easy, incidentally, to construct a numerical example in which the top left-hand submatrix has *all* the features of a prisoner's dilemma. This will be so, for instance, if the firms have identical cost curves. In this case, an in many ways reasonable solution of a 100-fold game of duopoly is for each to play the JM strategy until t (about 90), or until the other double-crosses. An objection (though not an unanswerable one) is that a series of duopolistic encounters does not normally have a fixed terminal date.

We finish our discussion of duopoly by noting briefly how Sweezy's well-known 'kinked demand curve' theory of oligopoly relates to all this. A Sweezy duopolist, like a Cournot one, attributes a reaction function to his rival. Here prices are the decision variables. Product differentiation means that prices may differ and yet both sell something. Given that both have been operating at some common conventional price, the conjectural variation function gives B's one-period-on responses to A's now raising, lowering or maintaining his own. B may react by himself setting his price high (H), low (L) or at the conventional level (C). The Sweezy reaction function is, in essence, (C, L, C), the three places corresponding to A's first choosing H, L, C, respectively. On plausible assumptions about cost and demand functions, ordinary profit maximisation leads A to continue to choose C. Similarly for B.

Like Cournot's model, Sweezy's differs from the game model in that (a) it is explicitly set in time. The 'reactions' will occur one period after their stimuli. In this they are like moves in an extensive-form game rather than simultaneously chosen game-long strategies. But, there is no true game, because (b) there is no uncertainty. The conjectural variation function tells the agent his opponent's current choice, which depends solely, and in a subjectively certain and hence irrational manner, on the agent's last one.

4.6 EXTERNALITIES [5]

Consider two firms which belong to a competitive industry, and which impose on each other external effects in production – that is, the production activity of each affects the production conditions of the other. The competitive assumption implies that the product price is given for the two firms. We shall express the external effects by saying that A's unit cost, the average cost of producing q_A, depends on both q_A and q_B; and that the average cost of q_B does so too. As in the duopoly game, both players are profit-maximisers, and risk plays no role and as in that game we may accordingly take utilities to be equal to profits. The model is again short run, and the decision variables are again the output levels q_A, q_B.

Game theory is needed because each firm's profit depends on what both do. Here the interdependence stems from the external effects. We shall see that these effects can give rise to two importantly different types of interdependence, a virulent strain, and a less virulent one.

Although A's profit depends on q_B as well as on q_A, this does not imply that A's profit-maximising output depends on q_B. That is, there may be a single output level, q_A^*, which maximises A's profit for every value of q_B. If this is the case, the external effect of q_B on A's problem only affects the size of his maximised profit, not the output level that maximises it. If there is such a q_A^*, an output level which is best whatever the other player's output level, then in game language this q_A^* is a strategy which dominates all other strategies of A's. A strategy like this which dominates all others is called *dominant*.

A competitor maximises his profit if he sets his output so that its marginal cost equals the given price. (We take for granted the subsidiary conditions – the 'second-order' condition, and the condition that price must cover average variable cost.) Firm A is a competitor. Hence, if A's marginal cost is independent of q_B, so must be his profit-maximising output.

Now A's marginal cost, denoted mc_A, is indeed independent of q_B if his total cost, denoted tc_A, has the form

$$tc_A = f(q_A) + g(q_B), \tag{4.3}$$

i.e. is the sum of two functions, the first involving only q_A and the second involving only q_B. For then mc_A, i.e. the partial derivative of tc_A with respect to q_A, is just the derivative of $f(q_A)$. Geometrically, a total cost curve of the form in equation 4.3 means that changes in B's output merely shift the curve showing the total cost of producing q_A up and down; but then they do not alter its slope, which is the marginal cost. If tc_A has the form in (4.3) it is said to be an 'additively separable' or simply a *separable* function of q_A and q_B. And also, in this case, the externality which consists of q_B's presence in A's cost function is called a *separable externality.*

The conclusion is that if the external effect on A is separable, then A has a dominant strategy: *separability implies domination.* In this case, though the externality is troublesome it is not intractable, as we shall now see.

4.7 EXTERNALITIES I: THE SEPARABLE CASE

EXAMPLE

The total cost functions that we used in the duopoly model of section 4.4 (cf. equation 4.1) were free of external effects. They were

$$\begin{aligned} tc_A &= c_A q_A = 4q_A - q_A^2 + q_A^3, \\ tc_B &= c_B q_B = 5q_B - q_B^2 + q_B^3, \end{aligned} \tag{4.4}$$

using equation 4.1. Now add a touch of reciprocal external effects to this model. Let us assume that both external effects are diseconomies, and specifically

$$\begin{aligned} tc_A &= 4q_A - q_A^2 + q_A^3 + \tfrac{3}{2}q_B^2, \\ tc_B &= 5q_B - q_B^2 + q_B^3 + \tfrac{3}{2}q_A^2. \end{aligned} \tag{4.5}$$

How will the payoff matrix look? In the no-externality case the payoffs to various values of q_A are independent of q_B and are straightforwardly obtained from the first of equations 4.4, once p is specified. With externalities, as in equations 4.5, each *column* of A's payoffs is simply the vector of no-externality payoffs for A's various output levels, with some constant added to every element of it, viz. $-q_B^2$. This constant is different for different columns. But because what is added is constant within each column, for each of B's strategies A's best counter is the same as before. So A still has a dominant strategy – and it is the same one as in the absence of externalities. Similarly for B.

Say $p = 6$. Consider only two strategies for each firm, one low and one high output level, say 1/2 and 1. Without externalities we have, from equations 4.4:

		q_B	
		1/2	1
q_A	1/2	(9/8, 5/8)	(9/8, 1)
	1	(2, 5/8)	(2, 1)

For each player the payoffs are independent of the other's strategy, and $(q_A^*, q_B^*) = (1, 1)$.

Now with the externalities. $q_B = 1/2$ subtracts $\frac{3}{2}(\frac{1}{2})^2 = \frac{3}{8}$ from A's payoffs. All the A components in column 1 must be 3/8 lower than before. Amending the whole matrix in a similar way, we get:

		q_B	
		1/2	1
q_A	1/2	(6/8, 2/8)	(−3/8, 5/8)
	1	(13/8, −7/8)	(4/8, −4/8)

Still, necessarily, $q_A = 1$ and $q_B = 1$ dominate. However, this strategy pair is no longer jointly maximal.

Indeed, we once again have a quasi prisoner's dilemma, like the top left-hand submatrix on p. 67. The two firms both stand to gain by slowing production to 1/2 each and so imposing much lower costs on each other – 3/8 instead of 12/8. In the present case it would not even be necessary for one to bribe the other: the strategy pair (1/2, 1/2) is not only jointly maximal but also Pareto-optimal. Indeed the only respect in which this reciprocal externalities game differs from the prisoner's dilemma of section 4.2 is that it is not symmetrical, and that is of no importance.

In the reciprocal externalities problem, as in the duopoly problem, the lack of cooperation leads to a 'jointly irrational' outcome. Here, moreover, there is a presumption that the outcome is irrational not only for the pair of protagonists, but also for the society in which their local activity is embedded. For if not only this industry but all others are competitive (and granted further conditions) money returns measure social value. *Society's* welfare would be increased if both cut back their outputs.

We have already suggested that the externalities in the present example are not a virulent strain. The separability of the external effects and the dominance property that goes with it make it possible to treat them by government policy, even though the firms continue not to cooperate. Provided that you know the payoff matrix, you can devise simple taxes (and/or subsidies) which will induce the spontaneous choice of (1/2, 1/2) by the non-cooperating firms. These taxes work by making (1/2, 1/2) dominant instead of (1, 1). They are taxes on the outputs of the two firms, and are simple in two ways: the tax on one firm's output does not involve the level of the other firm's output, and the rates are the same for both firms.

One tax that will have the desired effect is a tax of $-1+2q_K$ on firm K (K = A, B). (The lump-subsidy of 1 is to make the optimal outputs yield a positive and not just maximal profit, so that the desired outcome is viable in the long run.) With these taxes the payoffs become

		q_B	
		1/2	1
q_A	1/2	(6/8, 2/8)	(−3/8, −3/8)
	1	(5/8, −7/8)	(−4/8, −12/8)

[As an exercise, confirm that if the tax is to be a linear one $a+bq_K$, then (1/2, 1/2) becomes dominant, and profitable, as long as

$$a < -6/8, \qquad b > 14/8.]$$

Notice that because the taxes levied on each firm do not involve the other firm's output, they operate on the pattern of the payoffs in the matrix as 'internal effects': they alter A's columns in a way that is independent of B's output level. This makes it certain [why?] that each firm will still have a dominant strategy. Which strategies these will be is determined by the tax schedules. The preservation of dominance means that the government can be quite confident that the solution it wants will come about. It could not have this confidence if the post-tax payoffs rendered (1/2, 1/2) merely a non-cooperative equilibrium.

4.8 EXTERNALITIES II: THE NON-SEPARABLE CASE

EXAMPLE

Take the following pair of total cost functions:

$$tc_A = 4q_A - q_A^2 + q_A^3 q_B^2,$$

$$tc_B = 5q_B - q_A q_B^2 + q_B^3,$$

in which the external effects are no longer additively separable. With $p = 6$, we can compute:

		q_B		
		0·90	1·08	1·25
	1·0	(2·10, 0·98)	(1·92, *0·99*)	(**1·75**, 0·86)
q_A	1·16	(**2·26**, 1·11)	(**1·98**, *1·17*)	(1·71, 1·10)
	1·5	(2·21, 1·50)	(1·60, 1·58)	(1·03, *1·64*)

Here three strategies each are just enough to bring out the significant features of the complete payoff pattern.

An equilibrium pair of strategies is one whose payoff pair has a column maximum for A and a row maximum for B as its two components. In the above matrix, column maxima of u_A are shown in bold type and row maxima of u_B in italics. It can be seen that there is a unique equilibrium at the payoff pair (1·16, 1·08). But this equilibrium is *not* dominant.

If q_B is on the high side, this increases the weight of the q_A^3 term in tc_A: this in turn makes mc_A rise earlier and the profit-maximising q_A come out at a lower value. For example, for $q_B = 1{\cdot}25$ rather than 1·08, the best q_A is 1·0 rather than 1·16. Similarly, high q_A reinforces the effect of $-q_B^2$ in tc_B (which makes for decreasing marginal cost), and postpones the stage of increasing mc_B. For example, $q_A = 1{\cdot}5$ rather than 1·16 pushes up the best q_B from 1·08 to 1·25. In these ways the multiplicative form of the externalities produces the twisting visible in the pattern of the bold-type and italic entries, and destroys dominance.

Although there still is an equilibrium pair, it is in pure strategies, and it is unique, in the present example it is a much less eligible 'solution' than the one in the last example. There, as in all prisoners' dilemmas, it was without doubt the rational non-cooperative solution because it was dominant. Here it is no longer dominant. The equilibrium pair we have here also comes off badly by comparison with the equilibrium of a zero-sum (ZS) game. In ZS games equilibrium pairs are maximin (section 3.8). This is one of the properties that does not carry over to nonzero-sum (NZS) games. In NZS games it may or may not hold, and here it does not. A's maximin strategy is 1·0 rather than 1·16. Lastly, the equilibrium pair is not the joint maximum (this defect is one that is shared with equilibria in prisoners' dilemmas). (1·16, 1·08) is merely an equilibrium. It is nothing more.

We are forced to conclude that the solution of the non-separable externalities problem, like that of the duopoly problem, has not been determined by our theories. Game theory has thrown light on the problem, but it has not provided a definitive answer.

It should not, however, be too easily presumed that there is a definitive answer to be had. There is no warrant *a priori* for supposing that proper tests of individual and group rationality can be unambiguously defined, much less for assuming both that they can be defined and that a unique solution exists which passes them.

The 'twisted' form of the interdependence of the two firms not only robs us of a clear solution, but it also, as Davis and Whinston have argued [5], makes the policy problem intractable. At least, the classical Marshall–Meade taxes on the two outputs which worked in the separable case do not work now. These taxes can still shift the equilibrium pair to a socially better position; but they cannot make that pair dominant. Because the taxes are separate functions of the separate outputs, the contortion in the pattern of payoffs remains. The taxing ministry cannot therefore make sure that the shifted equilibrium will be chosen.

The joint interest of the pair of firms is to go to (2·26, 1·11), where the sum of their profits is maximal. At this point both could gain by A making a modest side-payment to B. Under generally competitive conditions, this output pair is also socially preferable. But if it is to be achieved, classical tax devices cannot be relied on, and it will have to be by merger, cartel or quantitative planning from above. Each of these paths to a cooperative solution has drawbacks. In particular, the first two pose a threat to the welfare advantages to which price-taking behaviour conduces, for the two firms taken together may no longer be insignificantly small.

These, then, are some of the consequences of independent decisions in an interdependent world.

4.9 OLIGOPOLY II: THE LIMIT SOLUTION

The theory of cooperative games, which takes up the second half of this book, depends on the concepts and results of the non-cooperative theory that we have developed over the last two chapters. But we have now almost finished with non-cooperative games as such. Before we take leave of them, however, we consider briefly a non-cooperative game for a large number of players.

Most of the basic properties of n-person non-cooperative games in which the number of players n is relatively small, have already figured in our two-person discussion. But what if n is large? And in particular, what happens to the Cournot/game solution for oligopoly if the number of firms in the industry is successively assumed to be larger and larger? In Chapter 7 we shall prove, by game-theoretic means, Edgeworth's famous 'limit theorem', which says that as the number of participants in an exchange market grows large, the only outcome that remains viable is a competitive equilibrium. In Edgeworth's model collaboration is possible between groups of individuals. Here we foreshadow this result in a purely non-cooperative setting.

Recall the duopoly solution which we called the 'efficient point'. It would be attained if each firm took the price as fixed and the price cleared the market. Now in Cournot's world each takes the other's *quantity* as fixed. But he does not take the price as fixed, since the price moves as his own output changes because of his not insignificant share of the market – and this, at least, he does understand. However, for large n the own-output effect on price tends to zero. Hence we might guess that the Cournot solution tends to the efficient point. Now the arguments of section 4.5, on the equivalence of the Cournot solution and the game equilibrium, work perfectly well, when suitably generalised, for an arbitrary finite number of firms n. So if the Cournot solution does tend to the efficient point, it follows that the game equilibrium does too. We now prove that the former proposition is indeed true.

**PROOF. There is a Cournot equilibrium if and only if

$$\frac{\partial}{\partial q_{\mathrm{K}}}(pq - c_{\mathrm{K}} q_{\mathrm{K}}) = 0 \qquad (\mathrm{K} = \mathrm{A}, \mathrm{B}, \ldots, \mathrm{N}), \tag{4.6}$$

where

$$p = \phi\left(\sum_{\mathrm{K}=\mathrm{A}}^{\mathrm{N}} q_{\mathrm{K}}\right),$$

$q_{\mathrm{K}} > 0$, and ϕ is the demand function. (Once again we ignore the short-run shutdown case of $\partial\pi_K/\partial q_{\mathrm{K}} < 0$, $q_{\mathrm{K}} = 0$, and assume second-order conditions satisfied.)

As n grows large – or rather, as we consider larger and larger values of n – the Kth firm's market share shrinks steadily. This is the cause of the dwindling own-output effect on price. It happens whether the total market size expands with n or not. The former is mathematically simpler. Specifically, we assume that the price at which any given total output can be sold in the n-firm case ($n = 1, 2, 3, \ldots$) is equal to the price at which $1/n$ of this output could be sold in the 1-firm case.

That is, if $\phi_{(1)}$, $\phi_{(n)}$ are the demand functions in the 1-firm and n-firm cases, we have

$$p = \phi_{(n)}\left(\sum_{\mathrm{A}}^{\mathrm{N}} q_{\mathrm{K}}\right) = \phi_{(1)}\left(\frac{1}{n}\sum_{\mathrm{A}}^{\mathrm{N}} q_{\mathrm{K}}\right).$$

Now

$$\frac{\partial p}{\partial q_K} = \frac{\partial \phi_{(n)}(\Sigma q_K)}{\partial q_K} = \frac{\partial \phi_{(1)}[(1/n)\Sigma q_K]}{\partial q_K} = \frac{1}{n}\frac{\partial \phi_{(1)}}{\partial q_K}.$$

Hence $\partial p/\partial q_K \to 0$ as $n \to \infty$. Thus, equation 4.6 implies that, in the limit, the Cournot equilibrium occurs where

$$p = \frac{d}{dq_K}(c_K q_K) = mc_K. \square **$$

5 Two-person Cooperative Games

5.1 INTRODUCTION

Suppose that the District Attorney blundered. Say the prisoners succeed in buying the services of a corrupt warder, or manage to tap messages to each other. The isolation which was the root cause of their terrible dilemma has been ended. There is no longer anything to prevent them from together resolving not to 'talk'.

But will the prisoners agree even now on mutual silence? Or rather, supposing they do, will they keep their word? There are the strongest reasons for each of them to welsh on such an agreement, to double-cross, and in the absence of some equally strong force backing the agreement, their promise of cooperation may prove empty. But *what* force – what can bring about true cooperation when, as here, short-run individual gain opposes it?

One answer is long-run individual gain, as in the 100-fold dilemma. Here I treat another as I wish to be treated by him in the future. A second answer is the existence of sanctions against default (ostracism, or worse, at the hands of the criminal fraternity). Both these motives for cooperation are, it will be noted, cases of enlightened self-interest, individually rational in some larger, but still purely non-cooperative game. Thirdly, a moral code may require it. It is an interesting question, but not one to be decided here, whether such a morality is ultimately analysable too as non-cooperative, or whether it is the mysterious artefact which marks the unanalysable transition from non-cooperative to cooperative human activities.

In any event, communication leading to worthless, frangible agreements is not enough. Cooperation entails something else if it is to be not merely promised but also realised – whatever that something may be. Game theorists, appreciating this need but impatient to get on and unwilling to track down the elusive something, merely ask: 'What outcome does rationality dictate if, in a game, there *is* the possibility of agreements on which strategy each player will follow, and if these agreements *are* sure to be kept?' A cooperative game, then, may be defined as one which admits of *preplay communication* to arrive at *binding agreements*.

There is, therefore, no question here of making disingenuous promises as manoeuvres in what is really a non-cooperative battle. You have to carry out what you promise. So if your promise is to be worth giving there must be real gains to be had from the *quid pro quo*. Not all games, therefore, offer scope for cooperation. Those do in which, roughly, both players stand to gain relatively to their non-cooperative prospects. The prisoners do, for they could achieve utilities of 9 each by (binding) agreement and they are condemned to 1 each if they pass up a chance to cooperate. The duopolists of the last chapter do: alone, they expect (say) the Cournot equilibrium payoffs; together they can get a higher joint return and both can end up the richer by a suitable distribution. But the players of a two-person *zero-sum* game have no reason to cooperate. The general insight given by the theory of cooperative games is that when a two-person decision is nonzero-sum there is an element of common interest. Cooperation helps, and is usually necessary, to exploit this common interest.

However, the realisation of the common interest depends also on an understanding about who will get how much of the potential joint return – about the *share-out*. Unless there can be agreement not only on strategies but also on this division of the fruits of joint action, joint action is unlikely to be seen. There is an identity of interest in bringing about one or another of the outcomes in a certain set; but there is a conflict of interest about which one. This conflict must be resolved, or else every point in the mutually desirable set may elude the players' grasp.

In cooperative games one characteristic feature of game theory – uncertainty – disappears. Once the players have agreed (I'll do this, you'll do that), they *know* their payoffs. (More exactly, since strategies may be randomised and the game may involve chance, they know them at least 'up to a probability distribution'.) Whatever problems identifying the 'solution' of a cooperative game may bring, finding a rational principle by which each player may hypothesise the other's strategy is not one of them. The latter may be regarded as the fundamental theoretical problem of non-cooperative games, produced by the coincidence there of interdependence and uncertainty. Here we have interdependence but open negotiation.

The old problem is replaced by the quite different, but as great, one of fixing on a principle for together picking out a pair of payoffs which is in some sense best for both. In a cooperative game there is usually a set of outcomes that are unequivocally worse than certain others, but of these latter, eligible ones, some will favour one player more, some the other player. These 'eligible' outcomes are those that satisfy the famous criterion of 'Pareto-optimality'. The question is, *which* of them is to be chosen? If the players fail to agree on this, the cooperative enterprise will founder. Can they perhaps pick out a particular eligible outcome by some principle of 'joint rationality'? Alas, it is easier said than done. The only unchallenged canon of joint rationality is the Pareto criterion itself, and all outcomes of the eligible set satisfy this equally. Thus cooperative game theory, like social choice theory in welfare economics, encounters an impasse in the notorious indeterminacy or 'incompleteness' of the rankings of outcomes given by the principle of Pareto-optimality.

Cooperative game theory, however, is not social choice theory under another name. In one interpretation of the latter theory the individuals are inescapably members of society, whether they like it or not; in another, paternalistic, interpretation, they are not regarded as decision-makers at all. In a cooperative game, on the other hand, a player is a very autonomous agent who always has as one of his strategies the option of contracting out, of withdrawing from the negotiations and falling back on his own resources. He will not lose his citizenship if he does this – merely the particular benefits on offer in the game. This has the immediate consequence that for each player there are bounds on the negotiable: the bounds are set by the payoff he could be sure to get if he signed no agreement and went his own way. These negotiation limits have no counterpart in social welfare theory. Notice that the limits within which agreements can be cooperatively negotiated are set by the payoffs to *non-cooperative* strategies. In this way cooperative theory rests upon non-cooperative theory.

So much by way of introduction.

5.2 THE ATTAINABLE SET

Any game in normal form is completely specified by a set of pure strategies for each player and the array of expected utilities or payoffs corresponding to various choices of strategies by the players. In a two-person game the payoffs may be shown in a matrix, in which the rows relate to one player's pure strategies and the columns to the other's. Each entry in the payoff matrix is a *pair* of payoffs (u_A, u_B). These payoff pairs (u_A, u_B) associated with the various strategy pairs may be plotted on a graph with u_A and u_B on the two axes. We shall say that this graph shows the 'space' of (i.e. all possible) payoff pairs, or simply that it shows 'payoff space'. Payoff space is the set on which most of two-person cooperative theory is enacted.

If each player has two pure strategies there will be four payoff pairs of this kind to be plotted; if player A has m and B has n pure strategies, there will be mn payoff pairs. To these points showing pairs of payoffs from different selections of pure strategies we must now add points given by one or both of the players using a *mixed* strategy. We then have the whole of the *non-cooperative attainable set*.

Cooperation enlarges the non-cooperative attainable set; it makes extra payoff pairs attainable. We now show this by an example.

THE BATTLE OF THE SEXES

Consider the payoff matrix (or payoff pair matrix)

		S	
		D	B
H	D	(2, 1)	(−1, −1)
	B	(−1, −1)	(1, 2)

She (S) prefers to go to the ballet (B), he (H) to the dogs (D); each prefers to go wherever it may be with the other, rather than go alone to his or her favourite spectacle. Suppose first that the game is played non-cooperatively. They have squabbled, and each must make his choice in ignorance of whether the other will turn up (but knowing, needless to say, the payoffs). The payoff pairs corresponding to the various pure strategy pairs are J, K and L (twice) in Figure 5.1.

Suppose she goes to the dogs and he randomises. Then his and her payoffs or expected utilities are given by the points composing the line LJ. For instance, if she plays D and he plays D and B with probabilities half and half, his payoff is $\frac{1}{2}\times 2+\frac{1}{2}\times -1 = \frac{1}{2}$, and her payoff is $\frac{1}{2}\times 1+\frac{1}{2}\times -1 = 0$ – which gives the payoff pair (1/2, 0), that is, the point N.

If only she randomises and he plays B for sure, they get payoff

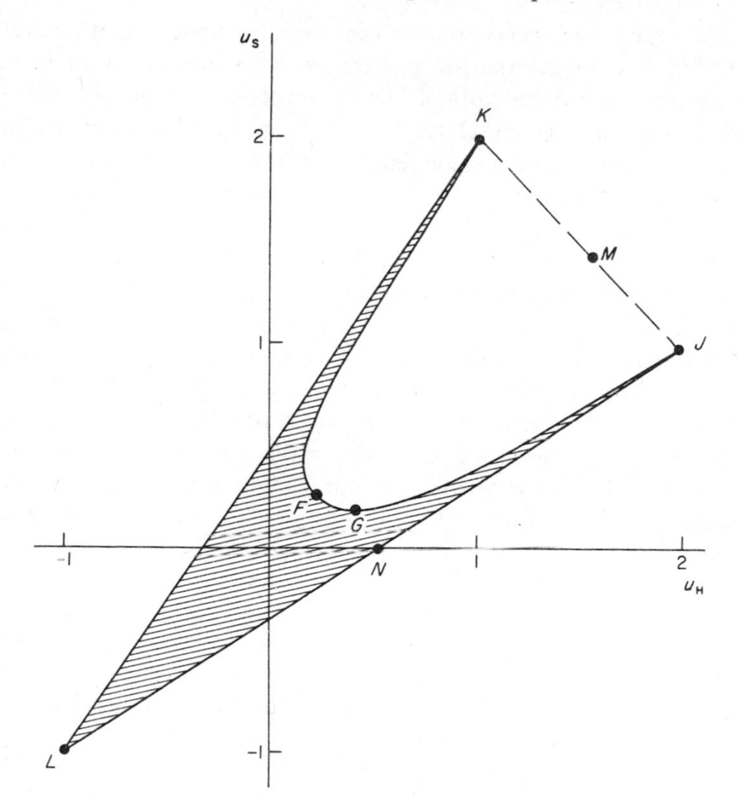

Fig. 5.1 *The battle of the sexes*

pairs along *LK*. [What is the locus of payoff pairs if she plays B and he randomises?]

Now suppose both players use mixed strategies. Say each tosses an unbiased coin and plays D if it comes up heads. Then each of the pure-strategy pairs (D, D), (D, B), (B, D), (B, B) has probability $(\frac{1}{2})^2 = \frac{1}{4}$. Hence H's expected utility is

$$\tfrac{1}{4} \times 2 + \tfrac{1}{4}(-1) + \tfrac{1}{4}(-1) + \tfrac{1}{4} \times 1 = \tfrac{1}{4}. \qquad (5.1)$$

So is S's, which establishes the point *F*. Similarly, if each has a random device (say a wheel) which yields some outcome with probability $\frac{2}{3}$, and plays D if and only if this outcome transpires, H's expected utility is

$$(\tfrac{1}{3})^2\, 2 + \tfrac{1}{3}.\tfrac{2}{3}(-1) + \tfrac{2}{3}.\tfrac{1}{3}(-1) + (\tfrac{2}{3})^2\, 1 = 5/9,$$

and S's is 2/9: the point *G*.

Generally, randomisations of this class, in which the probability of D is the same for the two players, generate the curved line *KFGJ* in payoff space. Finally, it can be shown that if the probabilities of the two players playing D are not equal, one gets the points making up the shaded interior of the figure *LKFGJL*.

We note two points before we go on. First, it is essential in the above expected-utility calculations that the two players' random experiments be *independent* of each other. Independence of the two randomisations is defined as meaning that the probability of one's getting a 'play D' result is unaffected by whether or not the other gets a 'play D' result. Independence is assured if the two experiments are not connected with each other either physically or by any passage of information. Secondly, remember that the payoffs to pure strategies – e.g. the payoff 2 to H for the pure pair (D, D) – may themselves be *expected* utilities; for chance (missed buses, indisposition of prima ballerina) may make moves in the game. However, one may take expected values of expected values by the usual arithmetic (multiplying each of the latter by the appropriate probability and summing),[1] so equation 5.1 is perfectly valid.

Adopt for a moment the long-run relative-frequency interpretation of mixing probabilities. Intuitively, if he and she have a series of dates (and they do not weary of their favourite spectacles, nor of each other), they should 'take turns' between (D, D) and (B, B). Now independent coin-tossing, or whatever, results sometimes in (D, B), (B, D); but *correlated* mixed strategies (which we now define) are free of this drawback. Suppose again that H plays D with probability $\frac{2}{3}$, the D-dates being decided by some random device or other, and suppose that S does the same. These mixed strategies are said to be (perfectly) correlated if the D-dates are the same in both of them. The correlation or matching is clearly impossible without some linkage between the two experiments: the wheels might be physically connected, or one player might simply be informed of the outcome of the other's experiment and match his play in this way. In the particular case in which the probability of D is 2/3,

$$u_A = \tfrac{2}{3}.2 + \tfrac{1}{3}.1 = 5/3 \quad \text{and} \quad u_B = \tfrac{2}{3}.1 + \tfrac{1}{3}.2 = 4/3,$$

and the payoff-pair $(u_A, u_B) = (5/3, 4/3)$, the point *M*.

---

[1] This assumes that the chance events in the game are probabilistically independent of the outcomes of the random experiments for mixing strategies.

The locus of payoff-pairs resulting from all perfectly correlated mixed strategies is the broken line KJ. Imperfectly correlated mixed strategies[1] give points within $KFJM$. We have shown that, in the Battle of the Sexes, cooperation extends the attainable region of the payoff space. We shall call this extended region the *attainable set* of the cooperative version of the game.

Cooperation extends the non-cooperative attainable set in a definite way which applies to any game. The eye shows that the extended region is the *convex hull* of the pure strategy points J, K, L. That is (see section 3.8 above), geometrically it is the region enclosed by a string drawn tightly round pins at J, K, L. Algebraically, it is the set of all *weighted averages* of the payoff-pairs corresponding to pure-strategy-pairs, e.g. the point $N = (\frac{1}{2}, 0) = \frac{1}{2}(2, 1) + \frac{1}{2}(-1, -1)$, the weighted average of $(2, 1)$ and $(-1, -1)$ with, in this case, *equal* weights. The origin $(0, 0) = \frac{2}{5}(2, 1) + \frac{2}{5}(1, 2) + \frac{1}{5}(-1, -1)$; the point $M = \frac{1}{3}(2, 1) + \frac{2}{3}(1, 2)$.

But then, equivalently, the convex hull of the pure-strategy payoff pairs is the set of all *probability mixtures* of pure-strategy payoff pairs; or, finally, it is the set of payoff pairs obtained by all probability mixtures of pure-strategy pairs. The rest is simple. By cooperation the players can achieve all such points, because they can mix *strategy pairs* in *any* way, perfectly correlated, imperfectly correlated, or not correlated at all. Without cooperation correlated mixtures are ruled out; hence without cooperation the attainable set is only a subset of the convex hull. In the present example this subset is the shaded region of Figure 5.1.

These considerations yield a neat formal definition of a cooperative game as one in which the set of attainable payoff pairs is the convex hull of (or equivalently, the set of all probability mixtures of) the payoff pairs of a given set of pure-strategy pairs.

Notice that in the non-cooperative version of the game even the points of the shaded area $LKFGJ$ are only 'attainable' in the sense that they *could* arise. But for point K, say, to be attained it would be necessary for the two players to estimate what was in each other's mind. K is merely a logically possible payoff; it is not attainable

---

[1] A's mixed strategy is imperfectly correlated with B's if A's playing D is made more probable by B's getting a 'play-D' result, without being made certain. It is clear that imperfectly as well as perfectly correlated strategies can be arranged by cooperation, and cannot be arranged without it.

in the sense that the pair of them can *decide* to to go K – it is not *choosable*. If the pair made a decision the game would, *ipso facto*, be cooperative, contrary to our assumption.

Thus the prospects that cooperation brings hold greater promise in two ways: first, the outcomes of a cooperative game are objects of the players' choice, while in a non-cooperative game nothing can be done to ensure that one rather than another comes about; secondly, there are more of them.

5.3 THE NEGOTIATION SET

In a cooperative game, then, the players can choose a payoff pair together; and their choice is wider than it would be if they could not communicate and correlate their strategies. We now turn to the question: *Which* point of the attainable set should the players choose? In other words, what is 'the' solution of a two-person cooperative game?

Two principles of rationality now *exclude* parts of the attainable set as acceptable solutions. In the case of two-person cooperative games these two principles do not generally narrow down the set to a unique point – the solution remains underdetermined. In n-person cooperative games, however, we shall see later that the same two principles can get us into the other trouble: they exclude so effectively that one may be left with *no* solution!

The first principle is *Pareto-optimality*. According to this principle the pair of players, if rational, will not choose a given payoff pair if there is another attainable one which has more expected utility for one player and at least as much for the other. Geometrically, the Pareto criterion restricts possible solutions to the 'north-east frontier' of the attainable set. This principle may be considered as one of the rationality of the two players considered as a *pair*.

The second principle is the 'go-it-alone' or 'get-stuffed' principle. Suppose player A can get expected utility v_A without cooperating; then if rational he will never accept a payoff pair (u_A, u_B) in which $u_A < v_A$. True, if a proposed cooperative solution gives such a u_A he might go along with it if he valued cooperation in itself. But there is no room for such an ethic within our framework, for in it the utilities already encapsulate *all* the values of the players.

The quantity v_A, the expected utility that A can reach on his own, is understood to be a cast-iron prospect, one that A can be sure of even if B should turn nasty; even, indeed, if B should cut off his own nose to spite A. It is also, evidently, the highest such secure payoff A can get by appropriately choosing his non-cooperative strategy. In other words, v_A is A's *maximin* payoff, or the maximum security level of all his strategies. We shall call this highest security level of all A's strategies the *security level of the game* for A – or simply, when there can be no confusion, A's *security level.*

This second principle may or may not eat into the north-east frontier to which the first principle says the players should confine their attention. In the Battle of the Sexes, his (and her) highest security level is -1. A glance at Figure 5.1 shows that all points of the north-east frontier *KJ* (the Pareto-optimal set) give both him and her considerably more than -1. So in this game the go-it-alone restriction does not bite.

Figure 5.2 represents a cooperative game in which the security levels do bite. *A,B,C,D,E* are the payoff pairs corresponding to the various pure-strategy pairs of the game, and *A,B,C,D,E* their convex hull, the attainable set of the cooperative game with these pure strategies. The pair of players can surely not rationally choose a point off the two-facet segment *ABC* (the north-east frontier or Pareto-optimal set.) But only points on the portion *FBG* pass the requirement that $u_A \geqslant v_A$, $u_B \geqslant v_B$. The set of points satisfying both principles (*FBG*) is known as the *negotiation set.* Von Neumann and Morgenstern say that any solution must lie in this set, which for this reason is also known as the *von Neumann and Morgenstern solution set.*

We see that the VNM solution of a two-person cooperative game is in general indeterminate. In particular, the essential indeterminacy of the Pareto criterion so familiar in welfare economics, although it has been mitigated by the requirement that $u_A \geqslant v_B$, $u_B \geqslant v_B$, has not been disposed of. Von Neumann and Morgenstern themselves thought that arguments from rationality could not identify a particular point as 'the' solution – though in any real game psychological and other factors extraneous to a theory of rational decision-making would often produce a determinate outcome.

We remark in anticipation that in the two-person cooperative game in which the two players are traders bartering goods, the negotiation set is the correlate in payoff space of Edgeworth's *contract curve* (which is a curve in 'goods space'.)

We shall be much concerned not only with games in which the

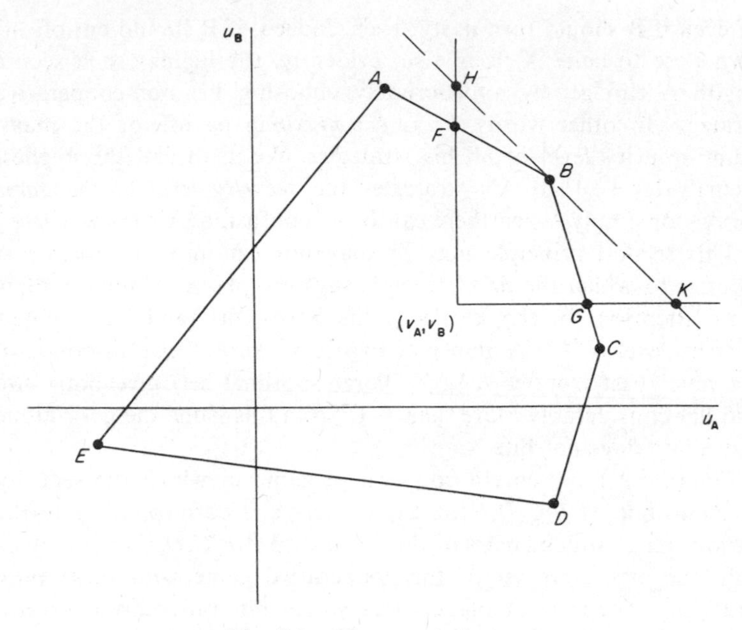

Fig. 5.2 *The negotiation set*

outcomes are exchanges of goods, but also with games in which the outcomes are monetary receipts. If the players' utilities are linear in money, then without loss of generality their utilities may be scaled and zeroed so that they may be numerically identified with their money receipts. Call a game like this a *game-for-money*. Suppose that, in a game-for-money, bribes or 'side payments' are within the rules – that is, the players may enter into agreements to redistribute their receipts after the play. It follows quite easily that the Pareto-optimal set is the 135° line through the most north-easterly point of the attainable set. It is the broken line in Figure 5.2; and the negotiation set becomes the segment *HK*.

5.4 BARGAINS

We have been left with a problem. How can one single out a point of the negotiation set as *the* solution of a cooperative game? There is a large literature which attempts to resolve the indeterminacy of the Pareto criterion, the notorious 'incompleteness' of the social preference-ordering which the Pareto criterion generates. The usual route, rejected for a long time but frequently taken more recently, is by interpersonal comparisons of utility. The attempt we now look at, due to Nash, game-theoretic rather than social choice-theoretic, eschews this route, and instead capitalises on the notion of the strategic strength that is conferred on an individual by his non-cooperative security level.

Consider a two-person game in which the strategies are decisions to trade commodities on various specific terms. Such a game is called a *game of exchange.* Suppose also that there is only one thing that can happen in the event of no trade being concluded. Then the game of exchange is called a *bargaining game* or sometimes simply a *bargain.* In a bargaining game there are no alternatives open to the players in their non-cooperative roles – in particular, there are no reprisals that they can take if their terms are not met. They just have to go home; at the end of the day they are still the owners of whatever they brought to market. The pair of payoffs if the outcome is 'no trade', if the players return home with their wares, is called the *status quo.* Note carefully that this term describes a point in payoff space, not a pair of baskets of goods.

A bargaining game [19] is fully specified by specifying:

(i) the cooperatively attainable region, R say, in payoff space, and

(ii) the *status quo* point, (s_A, s_B) say.

Points in R show the (expected) utilities the players would derive from settling with each other on these terms or that.

Notice that this definition of a bargaining game makes no mention of the goods that are traded. It therefore applies just as well if the outcomes are not 'trades', i.e. exchanges of commodities, but receipts of sums of money by each player say, or indeed any occurrences whatever that give rise to utility. Thus, any cooperative game may be regarded as a bargaining game provided only that the effect of a failure to agree sends the players back to a well-defined *status quo.*

It is clear now that the Battle of the Sexes was a bargaining game – a particularly simple one in which only a single unrandomised 'trade' is ever contemplated.

Next observe that the *status quo* payoffs s_A, s_B must be the players' security levels v_A, v_B in this cooperative game. Player A can be sure of s_A – this much expected utility is waiting for him at home; secondly, he cannot be sure of more, for whatever strategy he plays – whatever trade he offers, or should he offer none – B can hold him down to s_A by himself refusing to trade.

THE WASHING-UP GAME

Here is a simple bargaining game to illustrate these notions which features Him and Her some years later. The items to be traded are the preparing of (some or all) of the dinner and the washing-up of the dishes afterwards. His first strategy (α_1) is to sit back while she does the lot; under α_2, he proposes washing-up in exchange for her making dinner; in α_3, he offers to wash-up and prepare the salad while she cooks the rest of the meal. Her strategies β_1, β_2, β_3 consist in offering to carry out the trades reciprocal to those proposed in α_1, α_2, α_3. If they do not agree to one of these exchanges they will be forced to eat take-away Chinese food, a prospect we shall suppose he views with little enthusiasm and she with still less. Suppose too that he hates the thought of having to prepare the salad, and she the thought of his doing nothing. The payoffs might be as follows:

		S		
		β_1	β_2	β_3
	α_1	$(3, -1)$	$(1\frac{1}{2}, \frac{1}{2})$	$(1\frac{1}{2}, \frac{1}{2})$
H	α_2	$(1\frac{1}{2}, \frac{1}{2})$	$(2\frac{1}{2}, 1)$	$(1\frac{1}{2}, \frac{1}{2})$
	α_3	$(1\frac{1}{2}, \frac{1}{2})$	$(1\frac{1}{2}, \frac{1}{2})$	$(1, 2)$

The off-diagonal payoffs show the players' expected utilities if their demands are not acceded to. It makes no difference *which* demand is refused. If they fail to agree the outcome is always the same: take-away dinner. That is why the off-diagonal entries are all equal.

The attainable region R in payoff space and the *status quo* point (s_H, s_S) are as shown in Figure 5.3.

Note well that the notion that she dislikes the prospect of take-away Chinese food 'more than' he does is *not* expressed by her no-trade utility ($\frac{1}{2}$) being less than his ($1\frac{1}{2}$) – for the scales and origins of individual utility functions are arbitrary and *a fortiori* the utility indices of the two are not directly comparable. But the relation of his $1\frac{1}{2}$ to the 1, $2\frac{1}{2}$ or 3 he would have in other outcomes, when compared with the corresponding relation for her ($\frac{1}{2}$ in relation to -1, 1, 2) does express something about the degree of aversion with which each views the *status quo*.

It may seem unnatural to call the various trade offers the 'strategies' of the protagonists. It seems silly for H to contemplate 'playing' α_3 if S is willing to play β_2. Would not a better specification of his strategies have, say: 'If she is prepared to make the whole dinner, I'll not suggest I prepare the salad'? Where is the 'strategy' in artlessly revealing one's minimum terms at the very start?

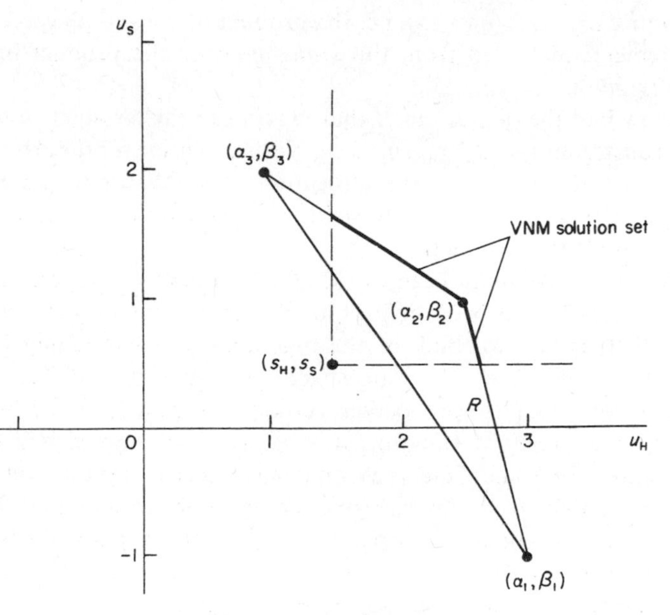

Fig. 5.3 *Washing-up*

The negotiations over the washing-up are indeed a very frank and open affair. All cards are on the table. The theory of bargaining games does not say what we might hear if we witnessed these negotiations – arguments based on interpersonal comparisons of utility, or on principles of fairness, displays of anger or tears, appeals to conventional standards? But it is clear what we would *not* observe: the dissembling manoeuvres typical of real bargaining, in which each party begins by exaggerating his true minimum terms, coming down only if necessary.

This is the inevitable consequence of a fundamental assumption of the theory of games – that each player knows the true preferences of the other, element (4) of section 3.1. It is perhaps in bargaining games that this assumption reaches its nadir of realism, for much of the flavour of real haggling comes from having to guess on what terms the other would be prepared to settle. But the theory sets out to describe not behaviour, but non-cooperative and co-operative modes of choice by rational beings aware of each other's interests.

Nash proposes the following formula for *arbitration* between recalcitrant bargainers. For any point (u_A, u_B) in R, consider the quantity $(u_A - s_A)(u_B - s_B)$, i.e. the product of the two players' utility increments measured from the *status quo*, or the product of their *utility gains.*

Now find the (u_A, u_B) in R that maximises this product subject to the constraints that $u_A \geqslant s_A, u_B \geqslant s_B$. Nash's solution for the arbitration of a bargaining game is the allocation of goods (or whatever it is that is being 'traded') which yields this (constrained) maximum of the utility-gains product.

Since we have so far been working mostly in payoff space, we must note carefully that Nash defines his bargaining solution in 'outcome space'. It is best to think of the outcomes of a bargaining game – points in the game's outcome space – as being pairs of baskets of goods, one going to each player. 'Good' is here interpreted broadly to include money, washing-up services; or anything which gives rise to utility. In some outcomes one may receive negative amounts of some goods – that is to say, one is a net supplier of them. We shall refer to the outcomes of cooperative games – these pairs of baskets – as *distributions.* It is clear that, just as there is an attainable or feasible set in payoff space so there is a feasible set of distributions in outcome space.

In passing, and largely for later reference, we define *demands.* A demand is a specification by a player of a basket of 'goods' for himself, i.e. his share of a distribution. It might be, say, (£50 per week, one new tea-vending machine, 42 hours per week, – 1), where the last item means the union gives the management a (unit of) undertaking not to submit a new claim for 12 months. There is a presumption that the player will settle if the other agrees to give him at least the quantities specified in his demand; for instance, to concede the union £50 per week and the tea machine in exchange for 41 hours' work and the 12-month undertaking. A pair of demands specifies a distribution. If the *distribution* is *feasible* (i.e. the players between them have it in their power to make available the total amounts of the various goods which it specifies) the *demands* are called *compatible.*

The Nash solution of the Washing-up game is (α_2, β_2) (she makes the whole dinner, he washes up) with probability 11/12 and (α_1, β_1) (she does the lot) with probability 1/12. The payoffs are $(2\frac{3}{8}, 1\frac{1}{12})$. Nash's claim that this is a good solution is made on the grounds that it satisfies the following four conditions which he holds to be desirable *a priori:*

(1) (Pareto-optimality.) A distribution should not be chosen if there is another distribution which is feasible and which one player prefers and the other does not disprefer.

(2) (Interpersonal non-comparability.) Suppose we re-scale or re-zero one individual's utility function; this should not change the solution. (It follows that re-scaling and/or re-zeroing both will not change it either.) We shall see the reason for the name of this condition in a minute.

(3) (Symmetry.) A bargaining game will be called *symmetric* if, in terms of *some* legitimate individual utility indices, R would be symmetric round the 45° line. If the game is symmetric, the solution should, says Nash, give the players equal payoffs in terms of the symmetrising utility indices.

(4) (Independence of irrelevant alternatives.) Suppose we remove certain trades from the game (say the government imposes a wage-freeze, or a trader is robbed of some of his goods). Then if the old solution *is* still available, it should still be chosen.

Condition **(1)**, Pareto-optimality, is obviously satisfied by Nash's solution, for if there were an outcome with payoffs u_A, u_B, one bigger and one as great as the payoffs of the Nash solution, then the latter would not maximise the product of utility gains, contrary to their definition. Note incidentally that the constraints on the maximisation ($u_A \geqslant s_A$, $u_B \geqslant s_B$) together with the fact that $s_A = v_A$, $s_B = v_B$, mean that the Nash payoffs constitute a VNM solution (are in the negotiation set): for the Nash solution both is Pareto-optimal and meets the players' security levels.

Condition **(2)** (interpersonal non-comparability) needs some explanation. It is often held that solutions of group choice problems should not depend on 'interpersonal comparisons of utility'. That is, they should not depend on the absolute amounts of utility ascribed to the two individuals, in the following sense. Considering any two outcomes, x, y, whether x or y is judged preferable for the group should depend *only* on whether or not A prefers x to y and whether or not B does, *not* on 'how much' each likes each. The idea behind this is the positivistic one that 'he dislikes this more than she likes that' is meaningless or at least unverifiable: according to this view sound prescriptions cannot be made on these shaky foundations. On the other hand, it appears that we do make precisely such judgements in family, corporate and political life and that a serious theory should allow room for such judgements, its negative role being limited to rooting out inconsistencies among them. Notice that banning interpersonal comparisons is not the same thing as denying unique determination of the individual utility index. One can construct a uniquely determined utility if there are a *summum bonum* and a *summum malum*, say eternal bliss (B) and perdition (P); moreover, we can do so for everyone, for these prospects are possibilities that everyone faces. By agreeing to write $u_K(B) = 1$, $u_K(P) = 0$ for each person K, one has a convention for scaling and zeroing which is universally applicable and well-determined. But this is a far cry from being able to say that a change in u_A from $\frac{1}{2}$ to 1 is equivalent to – would make up for – a change in u_B from $\frac{1}{2}$ to 0: from being able to say that one man's bliss compensates for another's damnation; neither this one-for-one trade-off nor any other is implied.

We now sketch a proof that Nash's solution satisfies condition **(2)**. Let u_A, u_B be any two utility indices for A, B. In terms of these indices, plot the set of attainable pairs of expected utilities net of the *status quo* utilities (in other words, draw the attainable set R on a graph with origin at (s_A, s_B); call these expected-utility gains or payoff gains g_A, g_B. The Nash solution is the outcome which maximises $g_A g_B$ – that is, it is at a tangency with the highest reachable 'indifference curve' of the form $g_A g_B =$ constant. Every such indifference curve is a square hyperbola and on it the elasticity of g_B with respect to g_A is -1 at every point. So the Nash point occurs where the elasticity of the utility-gains frontier is -1. (**This point is unique if the frontier is concave to the origin – but it is, since it is the frontier of R, merely shifted left and down by our subtraction of (s_A, s_B), and R is a convex set by its construction, so its north-east frontier is concave to the origin.**)

Now shift the origin of u_A to give an index u'_A, say. The utility gains are unchanged for every outcome, so the Nash outcome, where the elasticity of B's utility gain with respect to A's equals -1, is the same outcome as before. Again, change the units of u_A to give u''_A. But elasticity is a measure independent of units, so the elasticity still equals -1 for the same trade as before. This establishes the result.

Condition **(3)** (symmetry) is easy: one has only to notice that the family of square-hyperbola indifference curves is symmetric around the 45° line. *If* we can find u_A and u_B for which R is symmetric too (which maybe we can and maybe not), the tangency is clearly on this line. Note that, by condition **(2)**, we get the same *solution* (i.e. arbitrated exchange of utilogens) for any other u_A, u_B, although these will in general make R and the maximal

u_A, u_B asymmetric. This asymmetry is, as it were, merely an accident of perspective.

Condition (**4**) is called 'independence of irrelevant alternatives' by Marschak and by Luce and Raiffa. (Others, including Arrow [2], use 'independence of irrelevant alternatives' in another sense.) To see its meaning consider the analogy of rationing an ordinary (Hicksian) consumer's purchases. If the old chosen basket is within the rations it will still be chosen (at least if we assume that rations cannot be transferred to someone else or carried over). So the consumer's choices satisfy condition (**4**): but Nash's solution is determined, as we have seen, just like a Hicksian consumer's – by getting on to the farthest-out possible of a family of indifference curves. The situation is depicted in Figure 5.4. Condition (**4**) is not trite: unlike condition (**1**), it could be violated by a perfectly reasonable arbitration rule. For instance, *N* would *not* still be chosen by a rule that militated against 100 per cent fulfilment of one player's maximum potential gain.

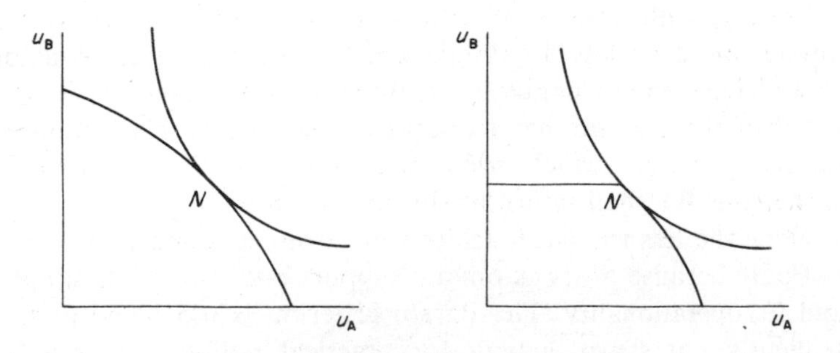

Fig. 5.4 *Independence of irrelevant alternatives*

Not only does Nash's formula satisfy conditions (**1**)–(**4**) but, more impressively, it is (it can be shown) the only one that does. However, these results by no means dispose of the problem we began with – how to single out one point of the negotiation set as the correct solution of a cooperative game. First, it has been put forward only as a suitable arbitration by a third party, and to expect disputants to accept an arbitration formula is not quite the same thing as to expect them to accept for themselves the solution to which this formula will lead. This objection is perhaps casuistic. It is psychologically relevant, but for rational agents the two cases arguably *are* equivalent.

Talk of arbitration is a bit of a front, or anyway a red herring, and Nash himself later abandons it.

Secondly, bargaining games are a subclass of cooperative games, special because of the very clear implications of a failure to agree: there is simply a reversion to a well-defined prior state. In more complex situations a player may be able to choose what the other has coming to him if he will not come to terms – i.e. he may dispose of a multiplicity of sanctions: these he may threaten the other with during the talks. In this more general case, which we consider in section 5.8, there is no *status quo* given *ex ante*. It will be decided (as we shall see) in a kind of non-cooperative 'threat game' which is part of the whole 'cooperative' game. Nash's formula is therefore inapplicable there. (It is true that in *any* cooperative game each player has a uniquely defined security level, his non-cooperative maximin gain v. The negotiation set as defined in section 5.3 is therefore always determined. But the *status quo* point (s_A, s_B) in a bargaining game is something more than the pair of security levels: it is the position to which the players *will* revert if there is no deal.)

Thirdly, with axioms of rationality, as with statistics, one can 'prove' more or less anything. The virtues and vices of axioms, as we have seen in Chapter 2, are often not transparent *a priori*. The proof of the pudding has to be in the eating. Nash's formula can produce answers which some may find questionable and even repugnant. We shall return to this in section 5.6.

None the less the Nash arbitration formula is appealing. It is so not only because of its axiomatic support but also for its simplicity and its operationality. The Pareto criterion is too bland and consequently not sharp enough for practical politics, but the Nash formula has the virtue of commitment: Nash says who should do the washing-up. (Herein lies another reason – a reason at the meta level – why the scheme should be accepted.) Problems remain: it is still impracticable whenever the utilities are unknown, and certainly the disputants have a motive for not telling the truth (in a larger, non-cooperative game with the arbitrator); but in monetary bargains for lowish stakes the arbitrator knows the utilities well enough.

5.5 ZEUTHEN'S BARGAINING THEORY

The Nash formula has a history. In 1930 Zeuthen [30] gave the Nash point as his solution for a descriptive theory of bargaining. Zeuthen models bargaining as a dynamic sequence of *concessions* by the two parties. Suppose that at some stage A is asking u_{AA} (more exactly, asking for a settlement in utilogens which yields him u_{AA}) and B is offering him u_{AB}. Similarly A is offering u_{BA} and B is asking u_{BB}, so that A and B are proposing the payoff pairs (u_{AA}, u_{BA}) and (u_{AB}, u_{BB}), respectively. A believes that if he holds out (niether concedes B's full demand u_{BB} nor makes any concession towards it) the possible consequences are: with probability r_A, an irreparable breakdown, in which case A gets his *status quo* utility s_A, and with probability $1-r_A$, concession by B of A's full claim. (Note that this implies that A assigns probability zero to B's making a partial concession as his next move, or else that we are talking of *conditional* probabilities that A assigns on the *hypothesis* that B will make no partial concession.) Then by the expected utility principle, A will do better to hang on than to agree to B's terms if and only if

$$r_A s_A + (1-r_A)u_{AA} < u_{AB},$$

or

$$r_A < \frac{u_{AA}-u_{AB}}{u_{AA}-s_A} = r_A^*, \quad \text{say.}$$

Symmetrically, B will prefer to hold out rather than to concede if and only if

$$r_B < \frac{u_{BB}-u_{BA}}{u_{BB}-s_B} = r_B^*, \quad \text{say.}$$

Zeuthen suggests that A will make a concession and B will not if and only if $r_A^* < r_B^*$; that is, if for A the probability-of-breakdown just sufficient to make total surrender preferable to total resistance, is less than it is for B. This suggestion seems arbitrary, especially in view of the skimpy treatment of the probability each assigns to the other's making a partial concession. But let us hear the rest, anyhow.

Without loss of generality we can put $s_A = s_B = 0$; i.e. measure the utilities of each from their *status quo* values. Then what we have is that A should concede if and only if

$$\frac{u_{AA}-u_{AB}}{u_{AA}} < \frac{u_{BB}-u_{BA}}{u_{BB}},$$

i.e. if and only if

$$(u_A u_B)_A < (u_A u_B)_B, \quad (5.2)$$

where $(u_A u_B)_A$ denotes the same thing as $u_{AA}\ u_{BA}$, that is, A's current proposal for the product of utility-gains, and similarly for $(u_A u_B)_B$.

Finally, suppose that if (5.2) is satisfied, so that it is A that should make a concession, this concession will be to propose a utility pair $(u_A' u_B')_A$ which will ensure that he will not have to make the next concession. For this it must be the case that

$$(u_A' u_B')_A > (u_A u_B)_B,$$

by a repetition of the argument that led up to (5.2). By reasoning similarly for B,

$$(u_A' u_B')_B > (u_A' u_B')_A,$$

and the successive new proposals $(u_A' u_B')_A$, $(u_A' u_B')_B$, $(u_A'' u_B'')_A$, $(u_A'' u_B'')_B$, . . ., for the utility-gains product will form an increasing sequence.

If the utility-gains frontier is concave to the origin (as cooperative theory tells us it must be) then this sequence must get nearer and nearer the Nash point: if the concessions do not become too niggardly too fast, it has the Nash point as its limit.

5.6 MORAL CONSIDERATIONS

It is not easy to decide whether Nash's scheme should be taken as technical guidance for professional arbitrators, as a prescription for *rational* decision-making by two people with nonzero-sum payoffs, or as an *ethical* prescription for the resolution of distributive conflict. As we suggested above, it may be possible to explicate at least some notions of interpersonal morality as types of group rationality, so the distinction between the second and third of these readings of Nash is not clear-cut. Nor is that between the first and the other two, since the good sense, or the justice, of an arbitrator's decisions may further his career.

If Nash's solution is 'good', it is so in a sense far removed from received Christian, or Socialist, doctrines. The Nash solution is *inegalitarian* in two ways. The first is necessary, built into the solution. The second depends on contingent psychological facts. The first, necessary, effect *preserves* existing inequalities. The second, contingent, effect, amplifies existing inequalities. The following examples show these two effects at work.

Both the examples concern two people, one of them, A, with a fortune of £10,000, and the other, B, penniless. By cooperating, they can make £1000 between them. (Let us say the profits depend on A's assets and B's brains.)

RICH MAN, POOR MAN – I

Let both their utilities be linear in money wealth, so without loss of generality we can put the utility of each equal to his wealth. The bargaining game is a game-for-money.

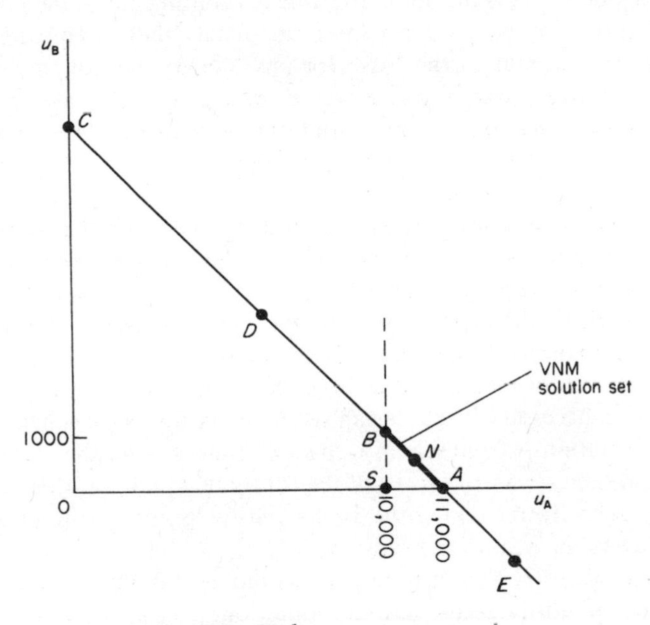

Fig. 5.5 *Rich man, poor man – I*

Point *S* in Figure 5.5 is the *status quo*. To draw the attainable set we need to specify the game a bit more. We have said that their cooperation will make them £1000 'between them'; this however is not an 'outcome' in the technical sense, since it does not specify who gets what if the two of them adopt such-and-such strategies. We must therefore construe our informal description of the outcome to mean that:

(i) the direct consequence of their cooperation is that A gets £x and B gets £$(1000-x)$, x being some number between 0 and 1000; and

(ii) the rules of the game – say, partnership law – allow 'side-payments' (redistributions of winnings) to be made freely between them after these direct consequences have come about. If this is the game, then the attainable pairs of payoffs from cooperation are given by the line *AB*.

This may not be all. It may be that redistributions of their initial fortunes are perfectly permissible too. If so, the attainable points after cooperation include the line segment *BC* got by A giving B more and more of the £10,000 be began with. Indeed if the game allows indebtedness the attainable line is prolonged into the negative quadrants! However, we can say immediately that these extensions of *AB* cannot contain the Nash (or any VNM) solution: points on it are non-negotiable, because they do not give each player at least his *status quo* utility (here, his initial money wealth). Player A would not cooperate if the price of cooperation were a point like *D*. Nor would B if it were a point like *E*.

Drawing new axes through *S* and considering the symmetry of the attainable set make it clear that in the Nash solution (marked *N*) each winds up with half of the extra £1000, so that the Nash solution yields the payoffs (10,500, 500). One cannot, you may say, say fairer than that. The solution is, it would seem, perfectly equitable in terms both of money and utilities. Yet this example was designed to show a necessary inegalitarian tendency in the Nash scheme.

The solution *is* equitable in terms of changes from the *status quo*. The *status quo*, however, is itself flagrantly unequal. So, therefore, is the solution if, not taking the *status quo* for granted, one measures the players' benefits 'from scratch'.

But if we took this last step we would be taking a liberty. The position of utility zeros, unlike money ones, is arbitrary – so that it is not clear that there is any such thing as measuring 'from scratch'. Even if it were, another obstacle would immediately loom up – interpersonal non-comparability. Zero utility for one man – being 'at scratch' – has to be shown to be a condition both comparable with, and similar to, zero utility for another, if there is to be any sense in saying that zero for both constitutes an 'equal' *status quo*. And this has not been shown. Consider again the Washing-up game. There, too, inequality of the *status quo* utilities ($1\frac{1}{2}$, $\frac{1}{2}$) contributed much to inequality in the utilities of the Nash solution measured on the same scales ($2\frac{3}{8}$, $1\frac{1}{12}$). We cannot say that his *status quo*, $1\frac{1}{2}$, the utility he would get from a take-away Chinese dinner, would leave him happier than her $\frac{1}{2}$ would leave her. The numbers are merely VNM utility assignments: one could add 1, or 100, to hers without making the least substantive change in the description of the state of affairs.

So jittery plutocrats can breathe again, agreeably reassured by the 'sound thinking' expressed in this sceptical objection.

On the other hand, if interpersonal comparability *is* admitted, the VNM interpretation of utility is left unharmed – indeed untouched. An *extra* coat of meaning is applied over it by a judicious assignment of relative utilities to the two individuals. Saying that the rich and poor man have *status quo* utilities of 10,000 and 0 is said in order to convey that one is contented and the other miserable. Because of the psychological meaningfulness of interpersonal comparisons which we have discussed before, we have no difficulty in reading the numbers in this sense.

If the utilogens are monetary, most people – and society – have strong ethical views on distribution. This may be because we are practised in contemplating the conditions, and consequences, of poverty and wealth. There seems little doubt that what we do so readily in the money case we could also do in others. And when we do it, interpersonal comparisons of the derived utilities, and not merely comparisons of the amounts of the objects going to the individuals, play a part in our judgements. Even in the Washing-up game, it is quite easy to 'read' the *status quo* utilities of $1\frac{1}{2}$, $\frac{1}{2}$ – and the utilities of the Nash solution of $2\frac{3}{8}$, $1\frac{1}{12}$ – in just this way. All this threatens the defence of Nash's conservation of inequality on grounds of non-comparability.

Let us agree that the Nash solution is inegalitarian in that it preserves pre-existing inequalities of *something*. (So, too, is every VNM solution?) These inequalities are often morally disquieting. The theory itself in no way justifies them. Indeed by the device of measuring from *status quo* utilities it neatly removes them from sight. This preservation of pre-existing inequalities is unexceptionable if one interprets Nash as merely saying that one *cannot expect* a bargain to be struck on any other terms, that only this distribution is viable, given the bargaining strength of the two parties. But if the solution is looked upon as an ethical prescription, as a contribution to social welfare theory, further defence is required; to accept the Nash solution without it is to concede that Might is Right.

The need for a defence is pointed up if we look once more at the Washing-up game. Whereas in Rich Man, Poor Man – I the preservation of initial inequality may strike most people as entirely 'natural', the same does not seem to be true of the Washing-up game. Why is this? What further defence can be offered for the conservative inegalitarianism of Nash's solution?

A half answer is that there are general attitudes in our society about what resources should be pooled in various kinds of joint enterprise. Marriage is one thing, business another. The ideal of conjugal sharing makes the Nash Washing-up solution grate on our moral sensibilities. But in the present game most would feel that A's and B's apparently *ad hoc* partnership imposes no such obligation. In short, Nash's solution will be accepted as *good* only when one feels that the players have 'property rights' in their *status quo* utilities.

We come now to the way in which, by virtue of a contingent psychological fact, Nash's solution increases inequality. The fact in question is risk aversion.

RICH MAN, POOR MAN – II [18]

Recall from section 2.5 the general empirical observation that VNM utility diminishes more sharply with income increments the bigger are these increments compared with the level of income. The same holds for wealth. That is, whereas a rich man like A might be indifferent between £500 and the lottery [$\frac{1}{2}$ £0, $\frac{1}{2}$ £1000], a poor man like B might have, say, £100 ~ [$\frac{1}{2}$ £0, $\frac{1}{2}$ £1000]. Accordingly, let us change Example 1 by curving B's utility function for money to express this risk aversion, leaving A's linear. We may also, for each of them, set the utilities of money gains of £0 and £1000 equal to 0 and 1, respectively, and thereby lose no generality within the VNM theory. Note that we make no interpersonal comparisons. We get:

money gain		*payoff*		$(u_A u_B)$
A	B	A	B	
0	1000	0	1·0	0
250	750	0·25	0·98	0·25
500	500	0·50	0·90	0·45
750	250	0·75	0·73	0·55
1000	0	1·0	0	0

The Nash solution gives £750 to A and £250 to B. Because there is risk aversion it has increased the pre-existing monetary inequality. It would do the reverse if the people had *convex* utility functions, i.e. if their slopes increased (the case of 'risk preference'.) Usually, as here, they do not; hence, generally, the Nash solution not only does nothing to redistribute progressively an unequal pre-game distribution of wealth, it redistributes regressively.

Notice, however, that each gets an approximately equal percentage of his maximum possible *utility* gain: B is soon satisfied by what is for him an exciting new experience; but A is already as it were heavily addicted to money. Hence, roughly equal percentage utility gains go with very unequal percentage money gains. It should be noted that the rough equality of percentage utility gains is not a universal property of Nash. It depends on the utility functions not being 'funnily' shaped (compare the situation depicted in the right-hand half of Figure 5.4 on p. 97).

5.7 WAGE DETERMINATION II [7]

We now have enough cooperative theory for a model of wage bargaining with real economic content. A *bilateral monopoly* is a market with one seller and one buyer. In many wage bargains a single union faces a single employer. The union (employer) could possibly sell (buy) labour elsewhere – to (from) another; but the negotiations which it is our aim to model do not involve third parties. These negotiations therefore constitute a two-'person' cooperative game. As long as we confine our attention to the direct relations between the parties, to this two-person game, we are describing – to use economic language – a bilateral monopoly. In the classical microeconomic literature the outcome in a bilateral monopoly – the quantity sold and the price – is held to be indeterminate. This indeterminacy turns out to be the familiar indeterminacy of the VNM solution set in disguise.

But if the two-person cooperative game is a bargaining game, the Nash settlement will provide a determinate solution that is rationalisable and not arbitrary. One of the notorious indeterminacies of classical microeconomics will thereby have been resolved.

For a two-person cooperative game to be a bargaining game there must be a unique outcome in case no agreement is struck. We shall assume this here. This means ignoring the variety of the actions the two sides can actually take in this case (strike, work-to-rule, overtime ban, lock-out, etc.) It allows us to use Nash's bargaining theory rather than the more difficult multiple-threat theory of the next section. (This latter does not always give a unique solution, although the VNM indeterminacy is much narrowed.) The main results are worth stating in advance.

(i) Output (and hence employment, since the model is short-run) is set as if the union and employer constituted a single enterprise buying labour from the external labour market and maximising profit.

(ii) The wage bill is determined by Nash's arbitration formula. 'No trade' is taken literally, to mean that *no* employment contract is arrived at, and production ceases. The union's share will be the greater, the higher its *status quo* utility – in effect, the larger its strike funds and other fall-back resources.

The model is as follows. The demand function is

$$p = \phi(q),$$

in the usual notation, and the production function is

$$q = f(e),$$

where e denotes employment. The firm's utility u_F is taken to be equal to profit π (the firm has no risk aversion). That is,

$$u_F = \pi = pq - we - rk,$$

where w denotes the wage-rate, k the capital stock employed and r the rental for capital.

The union's utility u_L (L is for labour) is some increasing function g of the 'wage surplus'. The wage surplus, S, is defined as the excess of the negotiated wage bill over the wages the same number of men could get in alternative occupations. Thus

$$u_L = g(S), \qquad S = we - w_0 e,$$

where w_0 denotes the 'external' wage-rate and g is an increasing function of S. This u_L has an advantage over those, like Dunlop's [8], which depend on we alone: for a union does not necessarily prefer to increase employment at the expense of the wage-rate merely because this would increase the total wage-bill of its members. Note that our u_L may be interpreted as the utility of a profit-seeking monopoly which transforms input (recruits) costing w_0 per unit into intermediate output selling at w per unit.

The first thing is to determine the negotiation set. It is important to realise that the negotiations are not just over w, but over w and e

(and hence, too, q and p). There is no convention or other prior restriction excluding certain matters from discussion; and both w and e are clearly pertinent issues, for together they determine the payoffs. Consequently, in this model there is not, *ab initio*, a well-defined labour demand curve giving e in terms of w as a function outside the union's control. (Such a demand curve is, certainly, appropriate in other types of market, e.g. in a competitive labour market, in which the sellers of labour do not bargain collectively with the employer, or in one in which they are unable to bargain with him over all the determinants of their payoffs.)

To determine the utility-possibility or Pareto frontier (of which the negotiation set or VNM solution set is a part) we first notice that

$$\pi + S = pq - w_0 e - rk.$$

This depends only on e (for e determines q, and q in turn determines p).

To get on to the utility-possibility frontier the two sides must agree to the level of employment e^* which maximises $\pi + S$. We show this by contradiction. Write $\pi + S = y$, and let e^* give $y = y^*$ (i.e. let y^* be the maximum value of y). Figure 5.6 is drawn in the space of money returns: its axes are π and S. The continuous 135° line in it is $\pi + S = y^*$. Choosing $e = e^*$ allows the players to reach any point on this line which they care to by a suitable distribution between them (and only such points.) If they choose some other

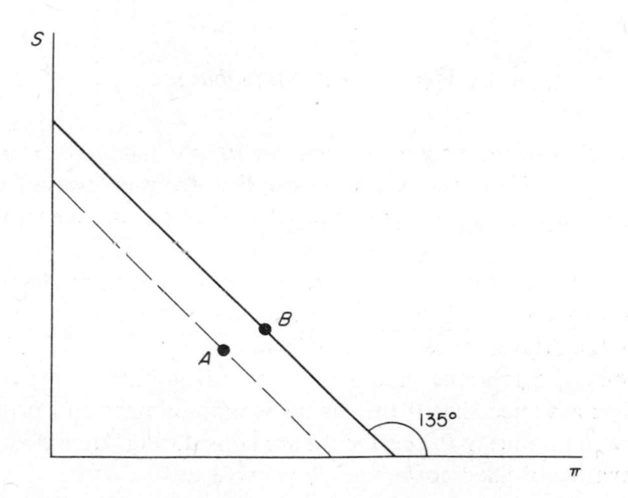

Fig. 5.6 *Feasible returns*

$e = e'$ which yields $\pi + S = y' < y^*$ they can reach (π, S) points along some line like the broken one, and only points along this line. But consider any point on the e' line – say A. Both π and S are higher at B on the e^* line. But u_F is *increasing* in π and u_L is *increasing* in S. Hence B yields higher payoffs to both parties. It follows that no payoff pair to which e' leads lies on the utility-possibility frontier.

Figure 5.7 superimposes (π, S)-space and (u_F, u_L)-space (payoff space.) The continuous 135° line is the same one as before.

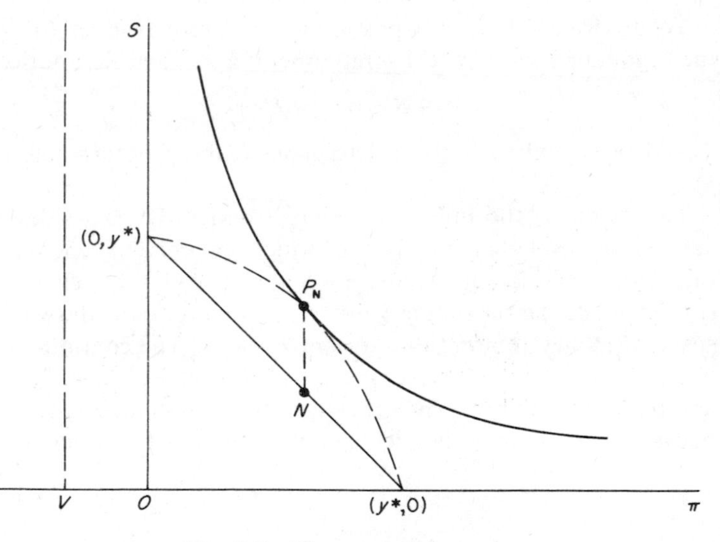

Fig. 5.7 *The attainable set*

Now choose the origin and unit of u_L so that $u_L(0) = g(0) = 0$, $u_L(y^*) = y^*$. Then the utility-possibility frontier passes through the endpoints $(y^*, 0)$, $(0, y^*)$ of this 135° line (its intercepts with the two axes). Finally, if, as we must suppose, u_L manifests positive risk aversion, the utility-possibility frontier must have the shape of the curved broken line. The region bounded by this curve and the two axes is the attainable set of payoff pairs.

Write π_0, S_0 for the values of π, S if negotiations break down. We have assumed that if this happens employment and production cease, at least during the period we are considering. Hence $\pi_0 = -rk$, the negative of fixed costs, and S_0 is given as

$$S_0 = (w - w_0)e = 0.$$

The utilities corresponding to π_0, S_0, i.e. v_F, v_L, must therefore also be $-rk$, 0, by our construction of the two utility functions. The points (π, S_0) and (v_F, v_L) therefore coincide in Figure 5.7, at V (though one belongs to returns space and the other to payoff space).

P_N is the Nash point in payoff space by the usual construction. Finally, N is the (π, S) point corresponding to P_N, i.e. N is Nash's *solution* of the bargain.

**We now determine N algebraically. From equation 5.2, for $\pi+S$, or y, to be maximised we need

$$\left(p+\frac{\mathrm{d}p}{\mathrm{d}q}q\right)\frac{\mathrm{d}q}{\mathrm{d}e} = w_0, \tag{5.3}$$

i.e. the marginal value product of labour equal to the 'external' wage-rate. Under the usual assumptions about the production and demand functions, equation 5.3 is also a sufficient condition and has a unique solution e^*. This determines the two frontiers (continuous and broken) in the last figure. To find N we need only put the elasticity of the union's utility increment with respect to the firm's equal to -1, that is

$$\frac{\mathrm{d}(u_L-v_L)}{\mathrm{d}(u_F-v_F)}\frac{u_F-v_F}{u_L-v_L} = -1.$$

Now $v_L = 0$, $u_F = \pi$, $v_F = -rk$, so this may be written

$$\frac{\mathrm{d}u_L}{\mathrm{d}\pi'}\frac{\pi'}{u_L} = -1,$$

where $\pi' = \pi+rk =$ profit gross of rental on capital. Then, equivalently,

$$\frac{\mathrm{d}u_L}{\mathrm{d}S}\frac{\mathrm{d}S}{\mathrm{d}\pi'}\frac{\pi'}{u_L} = -1.$$

But since the settlement is on the utility-possibility frontier, $\pi+S = y^*$, whence $\pi'+S = y^*+rk =$ constant, so $\mathrm{d}S/\mathrm{d}\pi' = -1$. Thus our condition becomes $(\mathrm{d}u_L/\mathrm{d}S)(\pi'/u_L) = 1$, or, since $\pi' = y^*+rk-S$,

$$\frac{\mathrm{d}u_L}{\mathrm{d}S}\frac{y^*+rk-S}{u_L} = 1.$$

Since u_L is given as a function $g(S)$ of S, this is an equation in

S alone. Once we have solved it for S, we shall have pinpointed the Nash settlement.**

In order to see the effect on the solution of different values of the firm's and the union's security levels, and of different degrees of risk aversion on the part of the union, we take $g(S)$ to be a specific form of function, namely a quadratic. Although quadratics have shortcomings as utility functions of money (see e.g. [3]), this will serve our present purpose well enough.

Write

$$u_L = g(S) = aS^2 + bS + c.$$

It is easy to check that choosing $u_L(0) = 0$, $u_L(y^*) = y^*$, as we have above, implies $c = 0$, $a = -(b-1)/y^*$, so that

$$u_L = -\frac{b-1}{y^*} S^2 + bS.$$

For this to be a plausible utility function, b must lie within certain limits. The marginal utility of the wage surplus is

$$\frac{du_L}{dS} = -2\frac{b-1}{y^*} S + b,$$

so we clearly need $b > 0$ in order that $du_L/dS > 0$ at $S = 0$, and $b > 1$ so that du_L/dS is diminishing – i.e. there is risk aversion. The greater is b the faster marginal utility diminishes, the sharper is the curvature of u_L and the greater is risk aversion. However, too high values of b would give $du_L/dS < 0$ as S increases. This certainly cannot be allowed for S between 0 and y^*, the range under consideration. Over this range, du_L/dS is a minimum at $S = y^*$, where it equals $2-b$. Thus b must be at most 2. We have, then,

$$1 \leqslant b \leqslant 2.$$

Set $y^* = 2$ (measured in, say, millions of pounds per month). Then we get the following Nash solutions for S as rk and b take the values shown.

		b		
		1	1·5	2
rk	0	1·0	0·90	0·85
	1	1·5	1·27	1·13

The figures in the first column exemplify that if the union had no risk aversion ($b = 1$), its wage surplus S, i.e. its absolute share of y^*, would be given by $\frac{1}{2}(y^*+rk) = \frac{1}{2}(y^*-\pi_0)$. This quantity also equals $\frac{1}{2}(y^*-\pi_0-S_0)$. That is, if neither side is risk averse, the sum of their potential money gains is divided equally. So too, by our utility scalings, is the sum of their potential utility gains. This illustrates the symmetry principle embodied in Nash's arbitration. A corollary of this result is that the greater are the firm's fixed costs – which is the same as saying, the lower is its security level – the more will the union get of the fixed total. We have already met both these properties of Nash in Rich Man, Poor Man – I in section 5.6, where we also discussed their ethical credentials.

Finally, given the potential joint money gain, y^*, the union's absolute share prescribed by Nash decreases with its risk aversion. This property was also discussed in section 5.6, with reference to Rich Man, Poor Man – II.

One footnote. It will be recalled from section 2.5 that, empirically, risk aversion decreases with absolute wealth or income. So the last property can be understood to mean that the worse off is the union (in the sense, say, of its members' social security benefits, its employed members' guaranteed wages, and its strike funds), the lower will be the wage settlement S.

5.8 NASH'S SOLUTION FOR THE GENERAL TWO-PERSON COOPERATIVE GAME

The cornerstone of the theory of bargaining games is the idea of the *status quo*. The *status quo* utilities are the players' security levels, and so they serve to narrow down the VNM solution set. Nash's determinate solution for bargains involves the use of a well-defined origin from which to measure utilities as increments, and it is the *status quo* point that serves this purpose too. Lastly, the use made of the *status quo* at each stage of the argument can be justified by appeal to non cooperative theory.

In this last section we describe Nash's theory of the general cooperative game [20]. His attempt to define a determinate rational outcome where VNM had admitted defeat depends on two ideas:

(i) All cooperative games are in the final analysis non-cooperative; there is always a latent non-cooperative game behind the cooperative goings-on.

(ii) A bargaining game is a cooperative game in which it is possible to define a determinate rational solution by exploiting the fact that in a bargaining game the strategies of the latent non-cooperative game are singular.

So a possible way to determinacy in the general cooperative game is: first, discover the latent non-cooperative game. Then, show this game to have unique *optimal* strategies. If this attempt is successful one will have shown that the general cooperative game has the same essential structure as a bargaining game, and the arguments to uniqueness in the bargaining situation will be applicable here.

'No trade' in a bargaining game may be thought of as an implicit 'threat'. Consider any course of action that a player in a cooperative game *could* carry out by himself, and that would in some way damage the other. Generally, a *threat* is a declaration by a player that he will carry out a course of actions of this type unless the other agrees to perform such and such a course of actions.

Now because a threatened action is an action to be carried out conditionally on the other's acting in a certain way – namely, promising to carry out certain actions – it cannot be a *strategy*. For a strategy is not conditional on anything (though its component moves certainly are). For example, in the Washing-up game, the threat not to share the chores in any manner does not appear at the head of any row or column of the payoff matrix (p. 92). This threatened action is, however, what lies behind and accounts for all the off-diagonal payoff pairs. Any of these payoff pairs shows what will result if the players' demands are incompatible, in which case there will be no agreement and consequently no sharing. Notice that in this game, as in every bargaining game, the threat does not have to be uttered. 'No-trade' gives each player a threat power that he cannot help exerting, that he has no need to flaunt before the other.

In Washing-up – and in every bargaining game – the outcome of a failure to agree is always the same. It is independent of the particular demands the players have put. Failure leads to 'no trade' *tout court*, a simple reversion to the *status quo*. Each player has this threat, and only this one. Bargaining games may therefore be classed as *fixed-threat* games. In the general cooperative game, on the other hand, each player disposes of alternative sanctions he may exercise in case of a breakdown of negotiations. Correspondingly he has a choice of threats, and so the positions in which the two players would find themselves if there were a breakdown is no longer necessarily the *status quo ante*. It is now a variable payoff pair, depending on what sanctions will be taken. Call it the *threat-point*. (Note that the threat-point is a point in payoff space, like the *status quo* in a fixed-threat game.)

So each player has a range of threats. It is clear that which threat he chooses to make is a matter he must decide for himself. Let us postpone consideration of how this choice might be made, and denote by t_A, t_B the threats that A, B (somehow) choose. These choices define a threat-point, the pair of payoffs that will result if t_A, t_B are carried out. Call these payoffs v_A, v_B. The threat-point (v_A, v_B) may now be regarded, formally, as the *status quo* of a bargaining game.

This *pseudo bargaining game* defined by a particular choice of threats is the key to Nash's solution of the general cooperative game. Since it is, formally, a bargaining game, it has a unique Nash solution. Consider adopting this as the solution of the whole cooperative game. We would have the uniqueness we seek. It would remain only to determine the solution of the 'threat game' – to single out optimal threats t_A and t_B. But the threat game is non-cooperative, so its solution should not create indeterminacy. These are the broad lines of Nash's strategy.

A critical and dubious step in it is taking the solution of the pseudo bargaining game defined by the chosen threats to be the solution of the whole game. Why should the players be disposed to accept the Nash settlement of this so-called 'bargaining game'? The rationality of their doing so is supported, it is true, by the list of axioms (1)–(4) in section 5.4. But it is not at all clear that it is *non*-cooperatively rational for them to accept these axioms. The rationale of the axioms lies in group choice theory, while we are here considering the players in non-cooperative, individualistic roles.

Nash therefore seeks purely non-cooperative arguments for the players to accept the Nash settlement of the pseudo bargaining game that results from their choice of threats. The arguments he adduces turn out to do this and more besides. They not only justify the Nash point given the threat pair, but in the same swoop they pick out a pair of optimal threats.

Consider the following two-move extensive-form non-cooperative game. At move one, A selects a threat t_A (and B a threat t_B). At move two, A decides on a *demand* d_A, and B on a demand d_B. If d_A and d_B are *compatible* – i.e. if (d_A, d_B) is a feasible distribution – the game ends, each actually meeting the other's demand. If they are not compatible, the game also ends, each carrying out his threat.

The demand d_A is allowed to depend on the opponent's threat t_B. This makes sense. An example of this kind of dependence is an employer's demanding less (offering more) if a union has warned that it will strike if its demands are not met rather than that it will work to rule. It also accords with the general model of an extensive-form game, where a later move in a complete strategy generally depends on the opponent's earlier moves. Indeed d_A should depend also on A's own earlier move t_A. So now what we have is a two-move non-cooperative game in which A's generic strategy has the form (t_A, d_A), d_A being a function of t_A and t_B; that is, a strategy for A has the form

$$(t_A, d_A(t_A, t_B)).$$

Every non-cooperative game has at least one equilibrium pair of (pure or mixed) strategies (cf. section 4.1.) Nash shows that the present game – the general multi-threat cooperative game metamorphosed into non-cooperative form – always has as one of its equilibrium pairs a pair $[(t_A^*, d_A^*), (t_B^*, d_B^*)]$ in which the demands d_A^*, d_B^* constitute the Nash solution of the bargaining game whose *status quo* is given by t_A^*, t_B^*!

Nash takes this starred equilibrium pair to be his solution of the cooperative game.

This is a *tour de force*. It is perhaps a meretricious one, for on the last lap it relies on the suspect claim that a non-cooperative equilibrium is a satisfactory non-cooperative solution. One trouble – among several – is that a non-cooperative equilibrium may not be unique. If there are several equilibria, it is not necessarily true of all of them that the demand pairs in them are the Nash settlements corresponding to the threat pairs in them. However, there is one important special class of multi-threat games, which we shall describe in a minute, in which the non-cooperative equilibrium is unique and the Nash proposal is altogether satisfying.

Nash's solution for the general, multi-threat, cooperative game has three striking features. We have already commented on two, namely that Nash solves the cooperative game by reducing it to non-cooperative form, and that he embodies his solution for bargaining games or fixed-threat cooperative games, for which he gives a different justification in the present context – non-cooperative equilibrium. The third striking feature is that threats are never carried out.

This should not shock us, since the model belongs to the theory of rationality, and is not intended to predict behaviour. That people do carry out threats – unions strike, countries go to war – does not mean that the courses of action which culminate thus are rational. Keeping this in mind, let us see why the players never do their damnedest here. The threats t_A, t_B are statements that A, B will carry out certain (damaging) acts if, and only if, their demands are not agreed to. The theory is not confined to games in which one is obliged to carry out one's threats (remember, these are moves in a *non-cooperative* game). Indeed the 'if' part of a threat statement might well be a deception (a bluff). But there is no reason why the 'only if' part should be, since neither player will make a demand that does not make him better off than would the implementation of his threat. So we can be confident that agreements – reciprocal acceptance of demands – if made, will be carried out.

Now in Nash's two-move extensive form there will be agreement if the two demands d_A, d_B are compatible, that is, if the amounts (of money, goods, services, etc.) which they specify do not add up to more than is available. But in any bargaining game Nash's settlement is feasible by its construction. Since d_A^*, d_B^* are the amounts in a Nash settlement they are compatible, hence there is agreement, and so it follows that t_A^*, t_B^* are not carried out.

We see now that the effective reason why threats are not realised in Nash is that at move two the players place 'Nash' and hence compatible demands. This is a little bit miraculous, for move two is part of a two-move non-cooperative strategy: d_A, for instance, has already been decided (as a function of the pair of threats) before A knows what d_B will be. But only a little bit: after all, the chosen d's are (part of) a pair of strategies in equilibrium, and as such are bound to have special properties.

If we imagine real players attempting to play the Nash solution we soon see why there are strikes, walk-outs, blow-ups, black eyes, and other breakdowns. There must be complete (which entails exact) information about the payoff matrix – which in turn requires complete information about the feasible outcome or distribution set, and about each other's utility function. For otherwise there is a danger of breakdown by mistake. Say A wishes to place the Nash demand d_A^*, and that this corresponds to the pseudo *status quo* consisting of the payoffs for t_A^*, t_B^*. Call these payoffs v_A^*, v_B^*. d_A^* depends on v_A^*, v_B^* because it is the constrained maximand of $(u_A^* - v_A^*)(u_B^* - v_B^*)$. So a mistaken estimate of v_B^*, say, will produce error in estimating d_A^*. If the latter error is positive it will tend to make the estimated d_A^* incompatible with B's demand (namely B's estimate of d_B^*.) This is what will happen if, say, A is an employer who underestimates v_B^* because he is unaware that the union can expect financial help from other unions in the event of a strike.

Thus, according to Nash's theory of cooperative games, strikes are the result of either irrationality or mistakes.

The problem of the possible non-uniqueness of non-cooperative equilibrium in Nash's threat-demand game disappears in the convenient case, which we have come across before (see p. 90), of money outcomes, linearity of utility functions, and legality of side-payments (the case of a game-for-money with side-payments). As usual, let v_A, v_B denote the pseudo *status quo* payoffs if the threats t_A, t_B are carried out. Here, we can think of these payoffs as (expected) money payments. The feasible set of pairs of payoffs – that is, money receipts – must be like the triangular area ODG in Figure 5.8. The origin O gives the game security levels of the two players; the north-east frontier has angle 135° (see Figure 5.2 on p. 90). If v_A, v_B were at the point V, the solution allocation would be given at N, where the 45° line through V intersects the utility-gains frontier EF. Hence (intuitively) A wants a V as far as possible in the direction of the arrow, and B one as far as possible in the opposite direction.

This suggests that a zero-sum game is being played over V. This is the case, as we now show. The line VN has equation $u_A - u_B =$

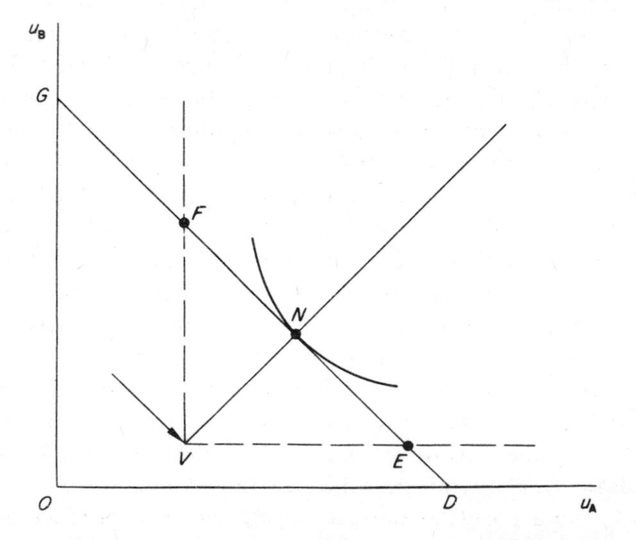

Fig. 5.8 *A multi-threat game-for-money with side-payments*

constant, the constant being $v_A - v_B$. The boundary EF has equation $u_A + u_B = \text{constant} = c$, say. Hence N is the point

$$\left[\frac{c}{2} + \frac{1}{2}(v_A - v_B), \frac{c}{2} - \frac{1}{2}(v_A - v_B)\right].$$

[Confirm this.] But this is the final outcome of the game if A plays the threat t_A and B plays the threat t_B. So the whole game is reduced to a non-cooperative game with strategies t_A, t_B and constant-sum payoffs. But it is easy to show that a constant-sum game can always be re-expressed, by appropriate choices of utility indices, as a zero-sum one. The upshot is that the whole game is unambiguously solved by the threats t_A^*, t_B^* which respectively maximin and minimax the payoff to A in this zero-sum game.

6 n-person Cooperative Games With Transferable Utility

6.1 COALITIONS

Going from the two case to the greater-than-two case is entering a different world. For as soon as there are three or more players in a cooperative game, some of them may form a coalition and play together against the remainder.

The concept of a coalition is that of a group of players who might decide to play as one. Formally, a *coalition* is merely a subset of the complete set of players. If there are n players, A, B, . . ., N, we write the complete set of players as $I_n = \{A, B, \ldots, N\}$. (The subscript n is just a reminder.) If S is a subset of I_n – that is, any set all of whose members are members of I_n – we write $S \subseteq I_n$ and S is a coalition.

Two things should be noticed about this formal definition. First, it says nothing about what the members of S will or will not do: it does not say that they *will* 'play as one'. In other words a coalition in the sense of the formal definition is something which it might be more natural to call a potential coalition. Secondly, S may consist of a single player, or of the whole of I_n. (To preclude the latter, we would have to have defined S as a *strict* subset of I_n and we would then have written $S \subset I_n$.) Thus, the pair of players in a two-person game is, formally speaking, a coalition.

A coalition is that which may act as a cooperative unit. Suppose that in a four-person game the coalition {A, B} does play together and that likewise {C, D} plays together. Playing together means cooperating in the sense of the last chapter: A and B can correlate mixed strategies, and they can promise each other (if the rules allow) recompense after the game has ended; their agreements are binding.

The coalition {A, B} may or may not now cooperate with the coalition {C, D}. The relations between these two coalitions have the same character as the relations between the two individuals in a two-person cooperative game like those of Chapter 5. Feelers may be put out between {A, B} and {C, D}, but neither will conclude a deal with the other unless this deal promises it more than it can be sure to get on its own. The notion of what a coalition can 'get' clearly requires definition, because the utility theory by which we have so far measured players' achievements is purely individualistic. Section 6.3 is devoted to answering this question.

A two-person game in normal form is fully specified by listing the pure strategies of the two players, and the payoff pairs which result from pairs of these pure strategies. Similarly, in an n-person cooperative game in which the 'players' may be coalitions, there must be a listing of the strategies which are available to coalitions as well as to individuals. We must know, too, the outcomes of combinations of strategies available to different possible combinations of coalitions. For example, in a three-person game with players A, B and C, what happens if A plays such-and-such, and the coalition {B, C} such-and-such?

The answer to this question will leave behind it another question – that of determining who gets what of the winnings which would accrue to a coalition if it played such-and-such a strategy. This, however, is not really a new problem for n-person cooperative theory, but one which is no different in essentials from the problem of determining the share-out in a two-person cooperative game.

6.2 TRANSFERABLE UTILITY

New vistas open out in n-person theory. Because which players will work together is not given in advance, the theory has a spicy, political flavour. There is something enjoyable about the shameless machinations which go on in the economic applications of the theory. One is very far from the slightly pi atmosphere of much neoclassical economics, in which one is seldom allowed to forget the social virtues claimed for competitive behaviour. Here by contrast we have an uncensored account of frankly rapacious economic behaviour.

For the next couple of sections, however, the new problems created by the variable line-ups of coalitions force us to submit the reader to a fairly long stretch of conceptual analysis.

It will be well to begin with some terminology for those cooperative games, of which we have already met some two-person examples, in which the game proper is followed by a further stage in which a final distribution of winnings occurs. Let us speak, whenever the distinction is necessary, of the *game proper* and the *share-out*. The game proper is an ordinary game in the sense defined at the start of

Chapter 3. The terms *strategy* and *outcome* pertain to the game proper. The share-out is a set of distributions (of goods, money, etc.), each of these distributions being a distribution within a coalition agreed by its members before play starts. There is no question of one coalition transacting with another after the game proper.

In some games, outcomes are goods, money, or other objects of utility which accrue to individuals, but in others the winnings accrue to coalitions as wholes, and before the second stage, the share-out, there are no personal winnings at all. Call these two kinds of games *incorporated* and *unincorporated.* If a game is unincorporated then the distribution which the share-out determines is a *re*distribution of winnings. These redistributions are called *side-payments.* In the usage of this book, side-payments do not have to be of money. Note that we can speak of side-payments in incorporated games too: there, once some provisional distribution has been made, or contemplated, any modification of it defines side-payments (relatively to this provisional distribution). One last term: let us call a player's final receipts (of various goods of money, etc.) his *cut*. A consequence of our definitions is that an unincorporated game with side-payments, a player's cut is equal to his winnings in the game proper plus any net side-payment he gets afterwards from the other members of his coalition.

Utility is said to be *transferable* between you and me if the actions open to us after the game proper has ended which result in changes in your utility produce opposite and *equal* changes in mine; that is, if the actions change, but conserve the sum of, our utilites. But this is precisely the situation if the game has monetary payoffs, side-payments of money are allowed and the utilities are linear in money. Remember, the units of our VNM utility functions can be chosen arbitrarily and hence can be chosen so that one pound has the same amount of VNM utility for us both. There can be transferable utility and yet many, indeed almost all, choices of utility units will render the sum of utilities non-constant. Utility is transferable if there is *some* choice of utility measures which renders them constant.

The work of VNM and their early successors which is the main subject of this chapter assumes transferable utility (TU) between all the players. It is helpful to bear in mind the money-linearity-side-payments interpretation. There is no need to imagine 'utility itself' being handed from person to person.

[Convince yourself, by considering the money-linearity-side-payments interpretation, that utility being transferable does not mean that it is interpersonally comparable.]

Notice carefully that TU does not imply that the game proper is zero-sum. It is only once the outcome corresponding to an n-tuple of strategies is established that total utility remains constant – over subsequent transfers of whatever gives rise to utility. All that we have here is a generalisation to the n-player case of the 135° line of utility possibilities in the two-person case: see Figure 5.2 (p. 90).

Cuts – what players will receive after intra-coalition distributions of winnings from the game proper – are offered explicitly during negotiations before the game proper, as inducements to join a coalition. We have already seen this in two-person cooperative games. In those games a player only agrees to cooperate if he is promised a big enough cut. But to cooperate is to work in the coalition consisting of both players.

In two-person cooperative games the basic requirements for the players' cuts to count as a solution were the VNM requirements – the conditions defining the negotiation set or VNM solution set. These, it will be recalled, are stated in terms of the utilities the players derive from whatever they finally get. More precisely, in terms of expected utilities, since the players' cuts or final receipts may be subject to risk. The requirements were:

(1) the pair of expected utilities, or payoffs, should be Pareto-optimal; and

(2) it should give each player at least his security level, i.e. his non-cooperative maximin payoff.

These requirements will have to be generalised. The generalisation of (1) to the n-case is obvious enough, but that of (2) is not. In a cooperative game with more than two players coalitions may form which act as single players. So to generalise (2) we have to define the security level of a coalition – what a coalition can be sure to get on its own. We do this in the next section. We shall see that it is the assumption of transferable utility which enables us to do it. Once we have successfully generalised (1) and (2), we will thereby have defined a generalised VNM solution set (or negotiation set). This generalised solution set is known as the *core*.

6.3 THE CHARACTERISTIC FUNCTION

Consider any coalition S which does not contain all the players, i.e. a set S such that not only $S \subseteq I_n$ but also $S \subset I_n$. Since we are assuming that utility is transferable, the utility that the coalition S receives in the outcome of the game proper is well-defined. (Think of utility as money.) This is a fundamental point which depends on the assumption of this chapter that utility is transferable. We may write this well-defined utility as $u(S)$. Now S cannot fare worse than if all the other players combine against it in a single rival coalition. This coalition of all the others, the set of players complementary to S, will be denoted by $\sim S$. The collaboration of all the other players in the single counter-coalition $\sim S$ is capable of damaging S more than any other line-up of the other players for the simple reason that $\sim S$ can do together anything that any breakdown of $\sim S$ can do – and possibly more things.

But if the players did coalesce into two groups S and $\sim S$ there would ensue a two-'person' game. Furthermore, it would be a two-person non-cooperative game – unless S and $\sim S$ changed their 'minds' and formed one grand coalition. Within this two-person non-cooperative game, let $v(S)$ denote the maximin value of $u(S)$. Then the quantity $v(S)$ is clearly the most that S can be sure of getting in the original game. So we have succeeded in defining the *security level* of an arbitrary coalition S (as long as S does not consist of all the players). Being able to do so depended on the assumption of TU, which allowed us to speak of the utility of a coalition. Notice that the definition of the security level of a coalition is quite consistent with our earlier definition of the security level of a single player in a two-person cooperative game.

How about the security level $v(I_n)$ of the coalition of all the players, the most they can get for sure if they all act together? For this all-inclusive coalition the adverbial clause 'for sure' is superfluous, because unlike strict-subset coalitions, I_n has no potential adversaries who might or might not use damaging counter-strategies. The environment is passive – the game is played undisturbed behind green baize doors. $v(I_n)$, then, is simply the most the group can get: that is, it is the maximum value of the sum of the utilities of all the players. In a two-person bargaining game-for-money, $v(I_n)$ is the maximum joint money return; in a zero-sum game $v(I_n) = 0$.

Now $v(S)$ may be regarded as a function defined on various coalitions S – it may in principle be evaluated for every $S \subseteq I_n$ in turn. In sections 6.5–6.7 we shall perform this evaluation in some concrete cases. $v(S)$ is called the *characteristic function* of the game. It gives the security levels of all coalitions, including, it may be noted, coalitions consisting of single players, and the grand coalition I_n.

Two properties of the characteristic function can be seen

immediately. The first is that if the n-person cooperative game happens to be zero-sum then $v(S)+v(\sim S)=0$. This comes from one of the corollaries of the minimax theorem (section 3.8) applied to the two-person non-cooperative game between S and $\sim S$. We remind the reader that S and $\sim S$ may have to use mixed strategies to achieve their maximin payoffs $v(S)$ and $v(\sim S)$. Whether they do or not, these maximin payoffs satisfy $v(S)+v(\sim S)=0$ by the corollary of the minimax theorem.

Now for the second property. Let R and S be any coalitions that are disjoint, that is, have no common members. $R \cup S$ denotes the set of players who belong to either R *or* S (it is the *union* of the sets R and S), and is itself a coalition and has a security level. The second property of v is that

$$v(R \cup S) \geqslant v(R)+v(S) \qquad (R,\ S \text{ disjoint}). \tag{6.1}$$

This follows from the consideration that acting together the members of R and S can achieve everything they can acting in two subgroups – and they can possibly achieve more. (6.1) says that, for disjoint groups, the function is *superadditive*. That is, the function of R plus S (i.e. of the union of R and S) equals or exceeds the function of R alone plus the function of S alone.

The characteristic function puts the cooperative game into a sort of reduced form. It records the results of the – often arduous – work of computing the maximin payoffs of all the possible coalitions there could be. We shall see as we go on that it is these maximin payoffs or security levels of different possible coalitions that define the rational outcomes of the n-person cooperative game: in particular, the security levels of different coalitions determine which coalitions different individuals will (if rational) join.

At this point we slip in another term. A game is called *essential* if some two coalitions can increase their joint security level by joining up. That is, a game is essential if there exist disjoint R and S such that $v(R \cup S) > v(R)+v(S)$. (Because of the superadditivity property (6.1), the only other possibility is equality.) R and S may of course be single individuals. If in a two-person bargaining game-for-money the utility-possibility frontier passes outside the *status quo* point, the bargaining game is essential. This follows easily from the fact that the security level $v(I_2)$ of the players as a pair is simply their maximum joint return – that is, it is the money level of the 135° straight-line utility-possibility frontier. By cooperating in this game, R and S can thus raise the sum of the receipts they can be sure of getting. Notice that if this is so, then rationality as expressed in the VNM solution set demands that R and S do cooperate – that the coalition that raises the joint security level does form.

A two-person zero-sum game is inessential: no improvement in joint security level is to be had from the only act of collusion that is possible, the cooperation of the two players. This comes from the fact that both $v(I_2)$ and the sum $v(A)+v(B)$ equal zero in a zero-sum game.

6.4 THE CORE

After the game proper, the share-out establishes an n-tuple of payoffs $(u_A, u_B, \ldots, u_N)$. We shall call this the *payoff vector* or simply the *payoff* of the n-person cooperative game. It reflects the *final outcome* of the game (as distinct from the *outcome*, which describes the end of the game proper).

We are now ready to define the generalised VNM solution set or *core*.[1] First, we have the Pareto-optimality condition. By some choice of strategies, the whole group can achieve a utility total of $v(I_n)$. In the share-out it can distribute $v(I_n)$ any way it chooses without breaking the rules. (Notice how the argument we are using depends on the idea of transferable utility.) It follows that if it agreed to strategies which resulted in a lower total payoff it would be agreeing to something on which it could improve in the Pareto sense. So we have:

(1) (Pareto-optimality.) The group must receive at least its security level, i.e.

$$\sum_{I_n} u_K \geqslant v(I_n). \tag{6.2}$$

Here $\sum_{I_n}$ denotes summation over all members K of the set I_n of players. Now by its definition, $v(I_n)$ is not only achievable but it is also the most that is achievable, that is,

$$\sum_{I_n} u_K \leqslant v(I_n). \tag{6.3}$$

---

[1] The core is not the only solution concept that has been put forward for n-person cooperative games (see e.g. [18]). In economic applications, however, it has received more attention than any other.

While equation (6.2) is a condition of rationality, (6.3) is a fact. Together, (6.2) and (6.3) imply

$$\sum_{I_n} u_K = v(I_n). \tag{6.4}$$

The second condition which a solution must satisfy is the straightforward generalisation of requirement (2) to n players: no individual should agree to a payoff vector in which he does not get at least his security level. No such payoff vector can be regarded as a solution. This gives:

(2) Each individual must receive at least his security level, i.e.

$$u_K \geqslant v(K), \qquad K = \text{A, B, \dots, N}. \tag{6.5}$$

We note in passing that a payoff vector which satisfies (6.4) and (6.5) is called an *imputation* in the literature. We shall however avoid using this term.

We come now to what is really new. If we impose the condition (6.3) that the whole group should not accept less than its security level, and the condition that no individual should accept less than his, by analogous reasoning we should require that

(3) Each coalition must receive at least its security level, i.e.

$$\sum_{S} u_K \geqslant v(S). \tag{6.6}$$

Suppose that the players are entertaining an agreement to play strategies and effect transfers which will result in the payoffs $(u'_{\text{A}}, u'_{\text{B}}, \dots, u'_N)$. If the sum of the u's that will go to members of the subset S is less than $v(S)$, the people in S should, if rational, jointly refuse to go along with what was about to take place. For by playing as a coalition they can be sure to achieve a bigger utility total, which then, by a suitable internal distribution, can be shared out to make *them* Pareto better-off than they would have been. The coalition S should therefore, in the jargon, *block* the payoff $(u'_{\text{A}}, u'_{\text{B}}, \dots, u'_N)$, that is to say, refuse to agree to it.

This argument is the lynch-pin of the theory of rational play in n-person cooperative games. The reader will recognise that the principle of rationality on which it depends – that a decision-unit should never accept less than its non-cooperative security level – is by no means new. All that is new is the application of this idea to decision-units of arbitrary size.

There is an apparent lacuna in the above argument to which we should draw attention. Where does the payoff proposal $(u'_{\text{A}}, u'_{\text{B}},$

$\ldots, u'_N$) come from? The simple answer is that it is suggested by the group acting as a whole – either in plenary session or through the agency of a secretariat. This is logically unexceptionable, but it is not very realistic if n is large. Notice that the implausibility of communication in the required degree of detail between a large number of people is just as much a criticism of (1) and (2) as of (3). In all three cases there has to *be* a proposed distribution to which to apply the tests.

It may be thought that the objection is irrelevant in the pure theory of rationality – that costs of communication are no more relevant to this theory than the imperfections of the human brain which we assumed away at the very start of the book. On the other hand, costs of communication could in principle be regarded as part of the outcome of pursuing certain strategies of negotiation, and could then be subsumed in the payoffs of a more complicated form of cooperative game. Some work has been done on these lines.

It is not in fact necessary to imagine that the group as a whole initiates proposals like ($u'_A, u'_B, \ldots, u'_N$). The alternative is to suppose that a proposed payoff comes out of a provisional coalition structure. The different coalitions composing this structure propose, in negotiations with each other, terms which will yield particular amounts of TU to each. At the same time, intra-coalition proposals divide up these amounts of TU to yield, finally, the utilities u'_A, u'_B, etc. to individuals. In short, the proposal ($u'_A, u'_B, \ldots, u'_N$) has come about by a devolution of the negotiations.

Notice that, because a coalition can include everyone, or only one person, condition (3) includes both conditions (1) and (2). A vector of payoffs which satisfies (3) and hence also (1) and (2), is said to be in the *core* of the game. The core is the n-person generalisation of the VNM solution set. The final outcome which corresponds to a payoff in the core is sometimes called a *cooperative equilibrium*.

Condition (3), which is only a natural generalisation of the conditions for the VNM solution set in a two-person cooperative game-for-money with side-payments, turns out to be most exigent. Whereas the VNM solution set, the two-person core, was too big – the solution was indeterminate – the n-person core may well be *empty*. That is to say, there may be *no* payoff that will not be blocked by some coalition.

On reflection this is not so surprising. Consider that the final outcomes whose payoffs are in the core are those which survive a most intense sort of competition between the players. Anyone may combine with anyone to block a distribution. There are no loyalties, no prior commitments, and the rules prohibit no patterns of collusion. Until everyone is satisfied, negotiations may continue during which players are free to overturn any arrangements they may have provisionally made. In another terminology, they are free to *recontract*. It is only in a technical sense that this kind of game is 'cooperative'!

It is quite easy to see mathematically why the core may contain no solutions. In the two-case the core is defined by three constraints – the Pareto or utility-possibility frontier and two side-conditions provided by the security levels of the two players. In the three-case there are seven constraints: the Pareto frontier; three side-conditions involving individual security levels; and in addition three side-conditions of the form $u_A + u_B \geqslant v(A, B)$. The number of conditions increases very fast as a function of n.

6.5 AN EMPTY CORE

We are now at last ready to look at another example. It is an example of a game which has no core. As a preliminary, we give a short theorem which specifies a class of games all of which must have empty cores.

THEOREM. If a zero-sum game for any number of players is essential, it has an empty core.

Alternatively, if a ZS game has a core, it is inessential: that is, if it is to have a solution in our sense then, in it, acts of cooperation necessarily fail to raise joint security levels. The proof takes this line, assuming a core and deriving inessentiality.

**PROOF. Let $(u_A, u_B, \ldots, u_N)$ be a payoff vector in the core. Then for any coalition S, both

$$\sum_{S} u_K \geqslant v(S) \tag{6.7}$$

and

$$\sum_{\sim S} u_K \geqslant v(\sim S). \tag{6.8}$$

The sum of the left-hand sides of (6.7) and (6.8) is just $\sum_{I_n} u_K$, which must equal zero since the game is zero-sum.

Suppose that (6.7) is a strict inequality. Then the sum of the left-hand sides of (6.7) and (6.8) is greater than $v(S)+v(\sim S)$. So in this case we have

$$0 > v(S)+v(\sim S).$$

However, by the minimax theorem applied to the two-'person' game between S and $\sim S$, $v(S)+v(\sim S) = 0$. The contradiction shows that (6.7) must be an equality:

$$\sum_S u_K = v(S).$$

This is true for any coalition. Consider any two which are disjoint, R, R' say. Then

$$v(R)+v(R') = \sum_R u_K + \sum_{R'} u_K = \sum_{R \cup R'} u_K = v(R \cup R').$$

That is, the game is inessential. □ **

SHARING A SHILLING

There are three players, A, B and C. For each of them one old penny gives one unit of utility. There is a shilling (12 old pence) on the table. The rules are simple. It is decided by majority vote which coalition gets the whole shilling. A little thought confirms the following values for the security levels of the different coalitions.

$$v(\mathrm{A}) = v(\mathrm{B}) = v(\mathrm{C}) = 0,$$

$$v(\mathrm{A, B}) = v(\mathrm{B, C}) = v(\mathrm{C, A}) = 12.$$

The game is certainly essential: A and B, for example, may achieve a security level of 12 by cooperating, while the sum of their security levels acting singly is zero. The theorem tells us that, being zero-sum, the game can therefore have no core, but let us check this. Suppose there is a core, and that $(u_{\mathrm{A}}, u_{\mathrm{B}}, u_{\mathrm{C}})$ is any payoff belonging to it. Then each of $u(\mathrm{A, B})$, $u(\mathrm{B, C})$, $u(\mathrm{C, A})$ is at least 12. Adding these three inequalities, we have $2u_{\mathrm{A}}+2u_{\mathrm{B}}+2u_{\mathrm{C}} \geqslant 36$, i.e. $u_{\mathrm{A}}+u_{\mathrm{B}}+u_{\mathrm{C}} \geqslant 18$. However, necessarily $u_{\mathrm{A}}+u_{\mathrm{B}}+u_{\mathrm{C}} \leqslant 12$ because there is only a shilling on the table.

What we have just given is a static non-existence proof by contradiction. It is analogous to showing the result that lowering wages cannot cure Keynesian unemployment by supposing that it does and showing that it follows logically that the interest rate must both go up and go down. Dynamics are more illuminating. Here is a glimpse of the shifting proposals and counter-proposals which might actually be made. The players are considering the distribution (4, 4, 4). Players B and C decide to combine to get the whole shilling. Their coalition {B, C} would not need to negotiate with A in order to do so, for 12 is its security level. As long as B and C stick together, the shilling is theirs. But will they? Say B and C agree internally on equal shares. If all of this went through, the payoff vector would be (0, 6, 6). But A now approaches C and, learning that he is to get 6 from this arrangement with B, tops B's offer: he offers C 7, which will still leave him with 5. This is not the end. Now, B can lure A away from C and still come out of the affair well. The 'recontracting' is endless. Every arrangement is frustrated. Every payoff vector is blocked. That is *why* the core is empty.

6.6 THE TRADING GAME

A *trading game* or *game of exchange* is a bargaining game, for n players ($n \geqslant 2$), in which the *status quo* corresponds to the baskets of goods which the trader-players bring to market. These baskets of goods are also called the *endowments* of the traders. The properties of these games furnish a highly developed and deeply satisfying theory of price, quite different in character and mode of analysis from, but quite consistent with, the supply–demand analysis that most people are brought up on. Monopoly and competition emerge naturally as limiting cases of a general theory. The rest of this book is about this theory. As in two-person bargaining, risk plays no explicit role.

There are two different approaches in the literature. One assumes, the other avoids, transferable utility. The TU subtheory, which we describe in the rest of this chapter, is simpler. It focuses on strategic questions about who goes with whom, and its racy numerical examples appeal to and sharpen one's sense of everything that is unscrupulous about market behaviour. The TU assumption is, however, by no means necessary for establishing the main results of n-person exchange theory.

The non-TU subtheory fits in better with modern mainstream microeconomics. It was developed by men who, unlike VNM, were 'normal' (in Kuhn's sense [15]) economists. The next chapter is devoted to it, but we mention one or two of its striking features here for comparison with the TU theory.

We need some new way, replacing a simple comparison of utility sums, to tell whether or not one coalition structure will supersede another. This is provided by standard neoclassical assumptions about individuals' preferences for commodities.

Although this later theory is non-TU, the share-out still plays a crucial role. All that changes is that it is the baskets of goods accruing to coalitions, or to their individual members, which get distributed, not transferable utility or money. Just as before, a player will refuse to cooperate in a coalition unless he is promised a big enough cut – enough of the various commodities.

The most celebrated result of the game theory of markets concerns the 'limit solution'. Economic intuition – and what we saw in section 4.9 about the oligopoly solution as n tends to infinity – suggest that the distribution of goods in the trading-game solution when n is very large may resemble a competitive equilibrium. But on second thoughts, would not this be quite remarkable? Oligopoly is a non-cooperative game. Might we not expect collusion – the essence of the present theory – to destroy the atomic property of competition, the helpless isolation of each competitor, and so to remove the cooperative limit solutions far from the solution of perfect competition? We shall see.

6.7 TRADING GAME I: ONE SELLER, TWO BUYERS, TRANSFERABLE UTILITY

In the two TU trading games which we describe in the next two sections only one commodity is traded, and it comes in indivisible units. Think of these units as houses of some standard specification.

In the first game a seller (A) has one unit to sell. There are two potential buyers B_1 and B_2, each of whom has no units – that is, neither has a house. A is willing to sell for a (units of money) or more. B_i ($i = 1, 2$) is willing to buy for b_i or less. Call a A's reservation of supply price, and b_i B_i's limit or demand price. All three players have utility linear in money. Put utility units equal to money units for each. Finally, choose the origin of the buyers' utility functions at their preplay positions, and the seller's at his preplay position *excluding* the house.

To solve this game – in other words, to determine who if anyone should buy the house and for how much – we evaluate the characteristic function; then we can determine the core.

$$v(\mathrm{A}) = a,$$

for the house is worth a to A, and $\{\mathrm{B}_1, \mathrm{B}_2\}$ *could* hold him down to this by refusing to buy.

$$v(\mathrm{B}_1) = 0,$$

since B_1 will make no gain above his preplay utility (*status quo*) if A, B_2 coalesce and make a deal which excludes him. Similarly,

$$v(\mathrm{B}_2) = 0.$$

We now tackle the trickier question of the security level of a coalition of one buyer and one seller. What is the most that A, B_1 can be sure of if they collaborate? They have the house. But even if B_2 values it highly he is not *obliged* to trade with them, there is no certainty that he will; A and B_1 may therefore be stuck with it. It follows that $v(\mathrm{A}, \mathrm{B}_1)$ must be whatever is the most they can achieve by internal dealings. Now if $a > b_1$, no such deal is on, and they finish (as a pair) with $a+0 = a$. If $a \leqslant b_1$, suppose that B_1 buys the house from A for a price p, where $a \leqslant p \leqslant b_1$. Then B_1 derives utility $b_1 - p$, and they finish with $p + (b_1 - p) = b_1$ between them. So we have shown that

$$v(\mathrm{A}, \mathrm{B}_1) = \max(a, b_1).$$

Similarly,

$$v(\mathrm{A}, \mathrm{B}_2) = \max(a, b_2).$$

And obviously,

$$v(\mathrm{B}_1, \mathrm{B}_2) = 0.$$

Lastly, we have to evaluate $v(\mathrm{A}, \mathrm{B}_1, \mathrm{B}_2) = v(I_3)$. By the 'superadditivity' property (equation 6.1), $v(\mathrm{A}, \mathrm{B}_1, \mathrm{B}_2) \geqslant v(\mathrm{A}, \mathrm{B}_1) + v(\mathrm{B}_2) = \max(a, b_1) + 0$, and likewise $(\mathrm{A}, \mathrm{B}_1, \mathrm{B}_2) \geqslant \max(a, b_2)$. Hence,

$$v(\mathrm{A}, \mathrm{B}_1, \mathrm{B}_2) \geqslant \max(a, b_1, b_2). \tag{6.9}$$

On the other hand, we now argue that

$$v(\mathrm{A}, \mathrm{B}_1, \mathrm{B}_2) \leqslant \max(a, b_1, b_2). \tag{6.10}$$

Only one person can finish with the house. The maximum utility sum achievable by virtue of final ownership of the *house* is thus max

(a, b_1, b_2). And, whatever *money* payments are made along the line in effecting the final outcome must cancel out since they are internal to the grand coalition. (This last point is only a more abstract version of the argument we used above for $v(\text{A}, \text{B}_1)$.) We have proved (6.10), and this together with (6.9) implies that

$$v(\text{A}, \text{B}_1, \text{B}_2) = \max (a, b_1, b_2).$$

Now for the core. Let (x, y_1, y_2) denote a payoff vector in the core. Then we can write down mechanically

$$x \geqslant a, \qquad y_1 \geqslant 0, \qquad y_2 \geqslant 0, \tag{6.11}$$

$$x + y_1 \geqslant \max (a, b_1), \quad x + y_2 \geqslant \max (a, b_2), \quad y_1 + y_2 \geqslant 0, \tag{6.12}$$

$$x + y_1 + y_2 = \max (a, b_1, b_2). \tag{6.13}$$

(Recall that we showed in section 6.4 (equation 6.4) that the relation between $\sum_{I_n} u_K$ and $v(I_n)$ is $=$ rather than $\geqslant$.)

This system of inequalities is easily solved. Consider the case in which $a < b_1, b_2$ and with negligible loss of generality let $b_1 < b_2$. So we are taking

$$a < b_1 < b_2. \tag{6.14}$$

Equation 6.13 implies

$$x + y_1 + y_2 = b_2, \tag{6.15}$$

but from (6.12) $x + y_2 \geqslant b_2$. Hence,

$$y_1 = 0. \tag{6.16}$$

Now because $y_2 \geqslant 0$ (6.11), equation 6.15 implies that

$$x \leqslant b_2. \tag{6.17}$$

But (6.12) implies that $x + y_1 \geqslant b_1$, and so

$$x \geqslant b_1. \tag{6.18}$$

(6.16)–(6.18) completely characterise the core, in the present case (6.14). x is A's terminal utility. Since $x \geqslant b_1$ it exceeds a, so A must have sold and x must be money: that is, x is the price A received. This price must (for the payoff vector to be in the core) be somewhere *between* b_1 and b_2, inclusive. That is as far as the core solution concept goes towards determining the market outcome.

The core solution is the same as the 'common-sense economics' solution. It corresponds to the band *GH* in the supply–demand diagram (Figure 6.1) – Böhm-Bawerk's 1891 solution for 'single-unit trade' [27]. But the supply–demand analysis is somewhat question-begging. In the familiar version in which buyers and sellers confront each other without the intervention *ex machina* of a Walrasian auctioneer, it is expressed in propositions of the form: if price p were offered by the demand side (resp. supply side) of the market, so much would be supplied (resp. demanded). But this glosses over the problem of saying what 'the demand side' means when, as here, it consists of two or more wheeling-dealing agents. It fails, for example, to come to terms with this question, very much to the point: could not B_1 and B_2 collude to get the house more cheaply, for less than b_1, splitting the gains from doing so by a side-payment from the front man, as happens in an auction ring?

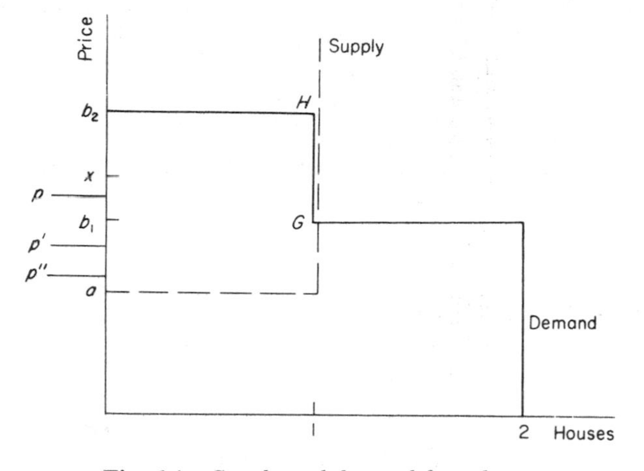

Fig. 6.1 *Supply and demand for a house*

We seem to have shown indirectly that such a deal would be wrecked, for a payoff in which both B_1 and B_2 gain over their *status quo* is not in the core (cf. equation 6.16). But let us look at the dynamics to find out how the ring gets broken. Let (x, y_1, y_2) be some core payoff or other about which B_1 and B_2 are having second thoughts. The distribution (x, y_1, y_2) should be thought of as being composed of two subdistributions: (x, y_2), being a distribution of $\max(a, b_2)$ agreed between A and B_2, and a 'distribution' of 0 over the 'coalition' consisting of B_1 alone. If B_2 is now to improve his position he must buy from B_1 for a price p' say $< x$; so B_1 must buy from A for $p'' < p'$. $\{B_1, B_2\}$ now propose to A that B_1 should pay A p'' for his house. Notice that $\{B_1, B_2\}$ can *not* enforce this arrangement, for p'' exceeds $v(B_1, B_2)$; they must put it to A. p'' exceeds A's security level; but this by itself does not satisfy A, who had just concluded an agreement with B_2 which would have given him $x > p''$. Player A sees that he can wreck the new deal by cutting out B_1, colluding with B_2 direct: he offers to sell to him for a price below p' but above p''. So the new proposed payoff does not succeed in ousting the old: $\{B_1, B_2\}$ lacks the non-cooperative muscle to block (x, y_1, y_2). [Why does the seller of a piece of antique furniture not get together with the dealer with the highest demand price as A does here with B_2?] [Can you see any way in which (x, y_1, y_2) could be blocked?]

The core solutions for market price are appropriate to conditions of unfettered market conflict with no holds barred and perfect information. They are determined by the static method in which one simply looks for values of the unknowns which satisfy a set of equations or inequalities. As we have now seen twice, however, we can also use the characteristic function to trace the dynamics, the contracting and recontracting which go on in reaching an equilibrium in these conditions.

6.8 TRADING GAME II: *m* SELLERS, *n* BUYERS, TRANSFERABLE UTILITY

Notice that we are making a small change in notational conventions: until now, n has denoted the number of players in a game, but here it is the number of buyers, the number of players being $m+n$.

The methods for determining the core in the case $m=2$, $n=2$ and for general m and n are just the same.

First let $m=2$, $n=2$ – there are, say, two houses for sale in the same terrace with owners A_1, A_2, and two potential buyers B_1, B_2 – and take the case $a_1 < a_2 < b_1 < b_2$ in an obvious notation. We shall only sketch the arguments which determine the core. See first that both houses must change hands, and for not more than b_1. If neither or only one were to change hands this would leave at least one buyer–seller pair who could both increase their utility by transacting. If the higher-priced one were about to sell for more than b_1, the other seller could undercut, making good business for himself and the man about to buy. Similar arguments show that both must sell for not less than a_2.

The novel feature when there are more than one unit traded is that all units must sell for exactly the same price. Here is a sketch of a proof. If (x_1, x_2, y_1, y_2) is in the core, then $x_1+x_2+y_1+y_2 = b_1+b_2$, b_1+b_2 being the maximum sum of pairs of the quantities a_1, a_2, b_1, b_2: this is the value of $v(I_4)$ for reasons analogous to those given in the last section. We also have $x_1+y_1 \geqslant b_1$, and $x_2+y_2 \geqslant b_2$. But since $x_1+x_2+y_1+y_2 = b_1+b_2$, both these inequalities must be equations, and in particular $x_1+y_1 = b_1$. In just the same way, $x_2+y_1 \geqslant b_1$ and $x_1+y_2 \geqslant b_2$ whence $x_2+y_1 = b_1$. It follows that $x_1 = x_2$.

In short, the only market outcome that is viable in the absence of binding prior commitments to particular coalitions – in the presence of unrestricted competition – is a common price somewhere in the interval between the highest supply price and the lowest demand price (inclusive). By drawing a diagram like the last one (Figure 6.1) it may be seen that such a price is given by the intersection of a jagged market supply curve and market demand curve. The core again gives exactly the same answer as Böhm-Bawerk's equilibrium for 'single-unit' trade.

We now let the numbers of sellers and buyers, m and n, be any numbers greater than or equal to 1. Without loss of generality we may suppose that $a_1 \leqslant a_2 \leqslant \ldots \leqslant a_m$, $b_1 \leqslant b_2 \leqslant \ldots \leqslant b_n$. We shall assume that $b_n > a_1$. Intuitively, this condition is necessary for any exchanges to be worthwhile. If it does not hold there is no possible feasible outcome involving exchange in which those who exchange improve on their no-trade positions. The players have nothing to say to each other. It is easy to show that, formally, $b_n > a_1$ is the condition for the game to be 'essential'. For example, Trading Game I is essential if and only if $\max(b_1, b_2) > a$, as it was in the case we considered (6.14).

It can be shown by the methods we have been using that the final owners are, for some number k, the k most eager buyers and the $m-k$ most reluctant sellers. The proof is mechanical but tedious, and we shall not go through it. All sales take place at a common price p, which lies somewhere between a_k (the highest supply price of those who do sell) and b_{n-k+1} (the lowest demand price of those who do buy). Figure 6.2 shows this schematically; here $k = 3$, and the square brackets indicate the supply and demand prices of those who transact; p lies somewhere in the interval $[a_3, b_{n-2}]$.

In spite of the restrictive assumptions which remain until we get rid of them in the next chapter – TU and indivisibility – this is an important result. We draw attention to four things about it.

First, different buyers get different payoffs, $b_n - p$, $b_{n-1} - p$, . . ., but all sellers get the same payoff p. This asymmetry is however deceptive, and simply comes from our choices of origin: if we had set sellers' *status quo* utilities at their money valuations of their property, namely a_1, a_2, . . ., rather than at zero, their payoffs would be measured as $p-a_1$, $p-a_2$, ... There is no implication that all the sellers are equally happy at the game's end. Neither does A_1 gain more happiness than A_2 just because $p-a_1>p-a_2$: the units of VNM utility functions are, of course, without meaning.

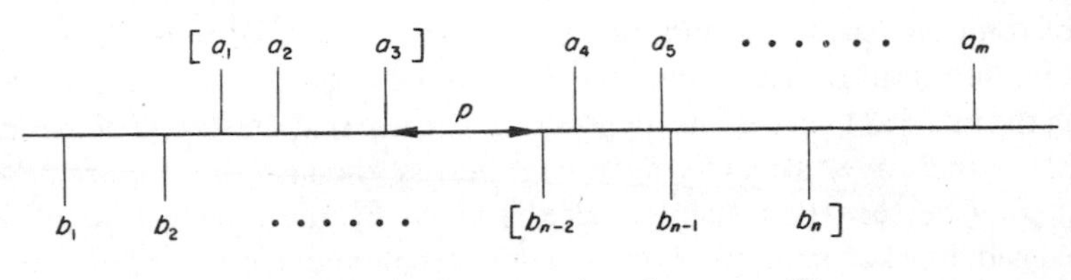

Fig. 6.2 *The core of trading game II*

It makes no difference who sells his house to whom.

Secondly, the state of affairs at the end is Pareto-optimal. Of course, it must be, for Pareto-optimality is one of the defining conditions of the core. We remind the reader that a payoff must be Pareto-optimal if it is not to be blocked by the coalition of all the players.

It may seem that the welfare properties of market outcomes are obtained with two little sweat in core theory. Pareto-optimality, the condition which competition theory labours so hard to establish, is simply assumed from the start. This view would be a misunderstanding. Core theory is part of game theory and game theory is a description of rational behaviour, both by individuals and groups. Pareto-optimality is a condition of rationality of the whole group, just as $\sum_S u_K \geqslant v(S)$ is one of the coalition S. Core theory therefore *begins* with rationality conditions, including the Pareto condition of group optimality; *its* labour is to establish the market outcomes which are implied by these conditions.

Thirdly, there is more than meets the eye in the assumption that each seller has only one house to sell. If a seller has two or more there is a possibility, unless we exclude it by a separate assumption, that his supply curve is backward-bending – or rather, because of the indivisibility of the units, backward-tacking: that is, there is a possibility that he will sell two at a lower unit price than he will sell one. Similarly, the possibility of upward-tacking individual demand curves cannot be – or from another point of view has to be – ruled out once we allow one man to buy two or more houses. This might be the shape of the demand curve of a property developer. Because we have excluded these possibilities by the rules, which restrict everyone to at most one unit, we have also made sure that the aggregate or market supply and demand curves are rising and falling respectively. Now in classical supply–demand theory these directions for these two curves are held to make equilibrium stable. The present theory provides independent confirmation of this claim. For we have seen that the classical Böhm-Bawerk equilibrium is in the core of Trading Games I and II, games whose rules imply 'properly' sloping market supply and demand curves. But to be in the core is precisely to be stable: no further manoeuvring would give a clear gain to any one, nor – a much stronger stability property – to any coalition. An arrangement in the core is one that will not be overthrown.

This, however, may be said of every arrangement in the core. Though none is, as it were, vulnerable from the outside, we cannot say that any configuration of transactions within the core is stronger than any other. We have not got rid of the old problem of indeterminacy. In short, stability of a solution does not imply uniqueness. The idea of a core has only enabled us to say of the market games we have studied up till now either that they have no solutions or that they have many. This is hardly satisfactory, although, as we have remarked before, it would be facile to conclude that what is unsatisfactory is the theory rather than the world.

The fourth and last property of Trading Game II provides some comfort. Consider larger and larger numbers of traders. As m and n grow, it is reasonable to assume that the spectrum of supply and demand prices will approach a continuum. If so, the indeterminate band in which the market price must lie – the gap between a_3 and b_{n-2} in the last example – becomes narrower and narrower, tending eventually to close up completely. As m and n tend to infinity, the core solution tends to one in which the price is not only common, and stable, but also determinate.

7 *n*-person Cooperative Games Without Transferable Utility

7.1 INTRODUCTION

As we have already intimated, all the main ideas of the last chapter except security levels of coalitions and the characteristic function easily survive the abandonment of transferable utility. Transferable utility made it particularly simple to say whether a coalition preferred one final outcome to another. But the essential criterion of group preference underlying it was always the Pareto one. If the inequality $\sum_S u_K > \sum_S u'_K$ was taken to signify that S preferred its payoffs in u to its payoffs in u', this was because its payoffs in u were redistributable so that for *each* individual K in S, $u_K \geqslant u'_K$, with $>$ rather than $=$ for at least one K of S. This underlying Pareto idea can be made to work equally well without (transferable utility).

In the last chapter there was in effect no distinction between utilities and objects of utility. Now that there is, it turns out to be convenient to define solutions in terms of final outcomes rather than the utilities or payoffs to which they give rise. All the n-person games we shall look at in this chapter are *trading games* or games of exchange; in all of them the final outcomes show how *commodities* are distributed among the players. The core will be taken to be a set in the space of these distributions of commodities. We could easily enough distinguish in our terminology between the core in the payoff space and this core in the space of final outcomes to which it corresponds, but using the same word is more natural, and it creates no confusion as we shall always be working in the latter space from now on.

A final outcome in an n-person trading game without TU specifies a *bundle* of goods for each player. Let x_K denote the bundle received by player K. x_K is a vector with one component for each of the different commodities that are being traded; e.g. if two goods are being traded it is a pair.

A *distribution* is a vector of bundles of goods, one bundle for each player. We shall write a distribution as $x = (x_A, x_B, \ldots, x_N)$. Thus a distribution is a vector whose elements are vectors.

Each player enters a trading game with a bundle, called his *endowment*. We write e_K for the Kth player's endowment. In Trading Game I of the last chapter, the endowments were 1, 0 and 0 for the three players A, B_1 and B_2. In general, however, each individual's endowment is not a single number but a vector, since there are several goods in play. The utility which his endowment gives to a player is his *status quo* utility.

If there is only one commodity, a distribution is *feasible* if it adds up to the sum of the endowments of the commodity. If there are several commodities it is feasible if the sum of the distributed amounts of the ith commodity equals the sum of the endowments of the ith commodity: this for each commodity i. Thus suppose there are two commodities, 1 and 2. Then x is feasible if

$$\begin{aligned} x_{A1}+\ldots+x_{N1} &= e_{A1}+\ldots+e_{N1}, \\ x_{A2}+\ldots+x_{N2} &= e_{A2}+\ldots+e_{N2}. \end{aligned} \tag{7.1}$$

This can be written more compactly in vector notation thus:

$$x_A+\ldots+x_N = e_A+\ldots+e_N. \tag{7.2}$$

We shall not need to use the cumbersome longhand of equation 7.1, and shall always use the compact notation of equation 7.2. The additions carried out on each side of (7.2) are known as *vector additions*, and each side of (7.2) as a *vector sum*. [Make sure you understand that the vector sum written $x_A+\ldots+x_N$ denotes the same thing as the vector

$$\begin{pmatrix} x_{A1}+\ldots+x_{N1} \\ x_{A2}+\ldots+x_{N2} \end{pmatrix}.]$$

Clearly the only distributions which can be final outcomes are feasible distributions.

The core of a TU game was the set of all payoffs $(u_A, \ldots, u_N)$ which passed the test $\sum_S u(K) \geqslant v(K)$ for every coalition S. Putting it another way, $(u_A, \ldots, u_N)$ was in the core unless this condition failed for at least one coalition S; if it did, that S would *block* it. The core was the set of unblockable payoff vectors. The underlying idea was that S would block a payoff vector if it offered S less than S could be sure to achieve on its own. We must re-express this idea in the new context, in which (a) the core is to be defined as a set of feasible distributions of goods rather than payoff vectors, and in which (b) we cannot simply express the desirability for S of the proposed final outcome, and of the best it can achieve on its own, as utility sums. There is little difficulty: what S can do on its own, and all that it can, is to redistribute internally (or of course leave unchanged) its own members' endowments. So all we need to say is: S will block a proposed distribution if some possible redistribution of its members' endowments would be preferable for it to the proposed distribution. And the meaning we shall attach to preferability for a coalition is the same as that which underlay last chapter's comparison of utility sums: Pareto-superiority. In sum, the members of S will block a proposed distribution x if by redistributing their endowments (or merely hanging on to them) they can make one member definitely better off and none worse off. If no coalition is disposed to block x for such reasons, x is in the core of the trading game. The core is the set of unblockable distributions.

In order to get concrete results about cores one has to make concrete assumptions about individual members' preferences among commodity bundles. We shall introduce these assumptions in the next section. They date back approximately to Hicks and are now standard in the theory of consumer demand. By adopting them we are implicitly treating our traders as final consumers and the commodities they exchange as consumer goods. This is intentional.

7.2 PREFERENCES AMONG COMMODITY BUNDLES

As in Chapter 2, $\gtrsim_K$ denotes 'weak preference' by K. That is, $x \gtrsim_K y$ means that K either prefers x to y or is indifferent between x and y. There is no harm in assuming that $\gtrsim_K$ satisfies the complete set of VNM axioms, but we do not in fact have to, because in all the games of this chapter the commodity bundles x which are the objects of preference are 'sure prospects'. We do need those parts of the VNM axiom set which relate to sure-prospect comparisons; i.e. we must assume that all pairs of bundles are comparable and that over trios of bundles the relation $\gtrsim_K$ is transitive. This we take as read.

The assumptions that are new are concrete ones about preferences for commodity bundles, not assumptions about rational choice in general.

(1) (Insatiability.) For any x, every $y > x$ is strictly preferred to x. That is,

$$y \succ_K x \quad \text{if} \quad y > x.$$

The notation $y > x$ means that the vector y is element-by-element greater than x.

(2) (Continuity.) Consider a sequence of bundles x^1, x^2, . . . approaching a 'limit bundle' x^*, i.e. the ith element of x^* is the limit of the ith elements of x^1, x^2, ... Then if x^1, x^2, . . . are all weakly preferred to a bundle y, x^* is weakly preferred to y. That is:

$$\text{If } x^1 \succsim_K y,\ x^2 \succsim_K y, \ldots, \text{ and } x^1, x^2, \ldots \rightarrow x^*, \text{ then } x^* \succsim_K y.$$

The meaning of this assumption is contained in a consequence which can be shown to follow from it. The consequence is: if a bundle x is strictly preferred to a bundle z, then any bundle x' which is close enough in all its elements to x is also strictly preferred to z. This corollary of assumption **(2)** is necessary for later arguments.

The condition is essentially technical and asks little. There is however one kind of situation where it breaks down, namely if K's preference ordering is *lexicographical*. $\succsim_K$ is lexicographical if: $x \succ_K y$ or $y \succ_K x$ according as x or y contains more of good 1; in case of equality, according as x or y contains more of good 2; etc. (The ordering of words in a dictionary is lexicographical – hence the name.) That this type of ordering violates assumption **(2)** can be seen by mulling over Figure 7.1. The last assumption is the key one for the theory.

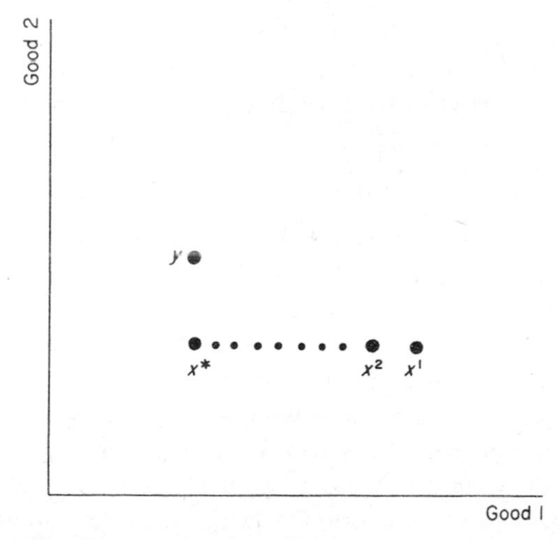

Fig. 7.1 *Continuity of consumers' preferences*

(3) (Convexity.) Let x and y be any two distinct bundles such that $x \succsim_K y$. If α is a number such that $0 < \alpha < 1$, then

$$\alpha x + (1 - \alpha) y \succ_K y. \tag{7.3}$$

The left-hand side of the preference relation 7.3 is a weighted average of the vectors or bundles x and y: its ith element is the fraction α of the ith element of x plus the fraction $1 - \alpha$ of the ith element of y. It is important in the sequel that the preference relation in the conclusion (7.3) is strict.

Assumption **(3)** implies, and is implied by, indifference curves which are strictly convex to the origin. (The qualification 'strictly' means that they do not have any linear segments.) So (7.3) may be thought of as a mathematical consequence of diminishing marginal rates of substitution between different commodities – or simply diminishing marginal utility (creeping surfeit) in each. For the record, a utility function which conforms with assumption **(3)** is called *strictly quasi-concave*.

7.3 TRADING GAME III: TWO TRADERS, TWO GOODS

The geometrical analysis we shall give is well-known outside game theory. It is partly for this reason that we give it – to show how economics textbook concepts such as *contract curve* correlate with those of game theory. Call the traders A and B, the goods 1 and 2. We assume (but only to make the diagrams cleaner) that each trader's endowment consists entirely of one of the two goods. Say

$$e_A = \begin{pmatrix} 140 \\ 0 \end{pmatrix}, \quad e_B = \begin{pmatrix} 0 \\ 120 \end{pmatrix}.$$

Each trader may offer to exchange whatever amount of his endowment he pleases in return for whatever amount he pleases. There are no indivisibilities or other restrictions on his trading strategies.

Figure 7.2 shows an Edgeworth box or Edgeworth Bowley box. $O_A D$ and $O_A E$ are axes showing amounts of goods 1 and 2 which A may finish up with. The upside-down back-to-front axes $O_B E$, $O_B D$ show amounts of goods 1 and 2 which B may get. For A one measures positive amounts away from O_A, for B, away from O_B. Point E is the 'endowment point' – both traders would end up at E if they made no exchanges. The curved lines filling the box are sets of indifference curves for A (continuous) and B (broken). These indifference curves satisfy assumptions **(1)**, **(2)** and **(3)** of the last section.

Consider any point in the box. Take its distances from $O_A E$ and $O_A D$ as amounts of 1 and 2 going to A, and its distances from $O_B D$ and $O_B E$ as amounts of 1 and 2 going to *B*. Then these amounts must make up a feasible distribution. [Satisfy yourself of this.]

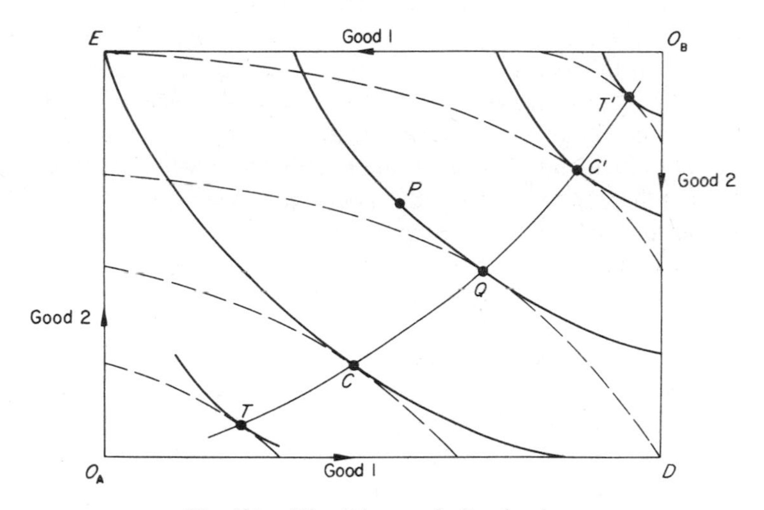

Fig. 7.2 *The Edgeworth–Bowley box*

The core has been defined as a subset of the set of feasible distributions. So it is some set of points in the Edgeworth box. There are only two traders, so what we have here is a two-person cooperative game (in fact, a bargaining game). Hence there are only two sets of restrictions defining the core: (i) Pareto-optimality and (ii) that individuals do no worse than by no-trade, for here no-trade gives them their security levels of utility.

It is a well-known result in textbook economics that for a distribution to be Pareto-optimal in this problem it must lie on the locus of common tangencies of the two sets of indifference curves – that is, on the curved line TT'; the figure shows why. Consider any point P such that the indifference curve of A that passes through it is not there tangent to an indifference curve of B. Then the point Q where this indifference curve of A intersects TT' must be Pareto-superior to P. For it is indifferent for A, and it is evident from looking at the figure that it is preferable for B. We do not make this optical argument rigorous, but it should be clear that a rigorous argument would appeal to the convexity assumption **(3)**, which determines the curvature of the indifferent curves, and to the insatiability assumption **(1)**, which implies that points on farther-out indifference curves are preferable.

So core points must lie on TT'. The security-level restrictions (ii) now rule out some of TT'. Any point on the T side of C, or on the T' side of C', gives a lower utility for one of the players than his endowment E, because it lies on a nearer-in indifference curve. The *core* in this two-trader game, therefore, is the segment CC' of TT'; CC' is also known in the literature as the *contract curve*. It will be noticed that our argument to the core is an argument about bartering which makes no mention of prices. Nor will it have escaped attention that the solution is indeterminate: all points of the core CC' are solutions according to game theory.

Now, to contrast with the core solution, consider what would happen if the two players were to behave like competitors; that is, if they were to maximise utility at some given and fixed prices for the two commodities. We can either imagine the two traders to be living in a competitive economy in which fixed market prices are given, or in a directed one in which prices are laid down by law.

In general there will be excess demand or excess supply of both commodities. Look at Figure 7.3. EF is drawn so that its slope is the given price ratio p_1/p_2. Then A can move from E to any point along EF by trading goods at the given prices, for at these prices p_2 units of good 1 exchange for p_1 units of good 2. Moreover, A cannot move except along EF. (He could if he had a money balance, or debt to sell, but in the present non-monetary economy his *only* endowments are 0 of good 1 and 120 of good 2.) For the same reasons, B too can move along EF, and only EF.

Utility maximising means that A will choose to be at the tangency point G, where he reaches the farthest-out indifference curve he can. B will choose to be at H. But because G and H do not coincide the desired baskets at G and H do not form a feasible distribution – their vector sum is not equal to the vector sum of the endowments. Hence, there is excess demand or supply of both commodities. The market is not in equilibrium. [Confirm that this is so by checking that the supply and demand of 1 are EG' and EH', respectively, and the supply and demand of 2 are EH'' and EG''.]

There is, however, in general one price ratio which will do the trick. We do not give the conditions for this to be so, but they are not exigent. In Figure 7.3 this price ratio is the slope of *EM*. The point where it touches an A indifference curve and the point where it touches a B indifference curve are the same point, *M*. The desired baskets coincide in the figure, so the desired baskets constitute a feasible distribution. Note that because *EM* is tangent to both indifference curves at the same point, the indifference curves must also touch each other at this point – the point must lie on *CC'*. So the equilibrium distribution is in the core.

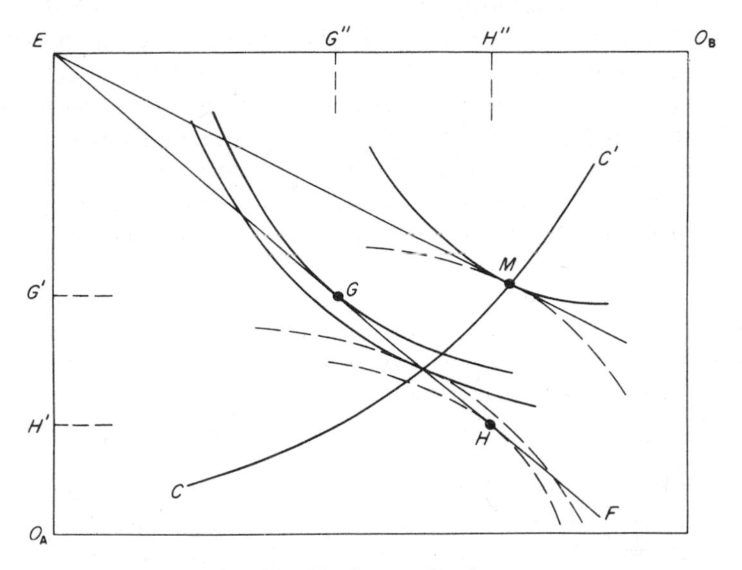

Fig. 7.3 *Trading at fixed prices*

Point *M* is called a *competitive equilibrium*, because price-taking is the prime feature of perfect competition. Geometrically it is the unique point of the core where the joint tangent of the indifference curves passes through the endowment point. It certainly gives a determinate answer to a question for once. But it can hardly be regarded as a solution of the two-trader problem we began with. Even if we are prepared, rather gratuitously, to embed the two traders in a competitive environment, we have given no reason why the price ratio that rules there should be that of *EM* rather than that of, say, *EF*.

Most of the rest of this final chapter is devoted to proving a famous result of microeconomic theory which says, very roughly, that in one class of *n*-person trading game a 'competitive equilibrium' analogous to M is the *only possible* solution. This loose statement needs three qualifications immediately. First, the class in question is the class of games in which n becomes indefinitely large – so there is no question of removing the indeterminacy of the solution of the last game, where $n = 2$. Secondly, the unique solution in question is a competitive equilibrium in the geometric sense in which we have defined M in terms of a joint tangent which passes through E: no appeal whatsoever is made to prices somehow exogenously given to the participants. Thirdly, it is the only 'possible' solution in the sense of core theory, the only distribution which would not be blocked by some coalition; that is, it is the only point in the core.

It is these last two points that give this result its great importance and put it on a par with the two other fundamental theorems in general equilibrium analysis. Walrasian theory showed that a competitive equilibrium exists: there *are*, under certain conditions, prices such that *if* all agents take them as given, maximising behaviour produces a universal balance of supply and demand. Paretian theory showed that an allocation of resources is *optimal* if and only if it is a competitive equilibrium in Walras's sense. The present theory, pioneered by Edgeworth, shows that in a ruthlessly mercenary collusive regime with many agents the only *possible* allocation – the only one which will not be wrecked – is a competitive equilibrium. What is remarkable is that the outcome in a Walrasian economy in which helpless little men with no telephones passively accept the prices to which the omnipotent invisible finger points, and the outcome in the Edgeworthian *casbah* in which ten thousand sharp operators scurry about fixing things up, are exactly the same. Deals in the limit are like no deals at all.

7.4 TRADING GAME IV: TWO TYPES, TWO GOODS

The ingenious analytical move which cracks the many-trader problem is to replicate the original traders A and B of the last game. Rather than considering many traders with any old tastes and endowments we consider many A-like traders and many B-like ones. This is certainly an implausible model even as an approximation. The heuristic is typical of economics. It allows us first to discover as it were the pure effect of number. Later, one can generalise to arbitrary many traders of arbitrarily many types. This has in fact been done [6].

We begin by just duplicating A and B. Thus A_1 and A_2 have identical tastes described by a preference relation $\succsim_A$ and identical endowments e_A. B_1 and B_2 both have preferences $\succsim_B$ and endowments e_B. We first prove the

PARITY THEOREM. If a distribution is in the core it assigns the same commodity bundle to both traders of a given type.

PROOF. We show that a distribution which does not do this is *not* in the core. Consider any feasible distribution $x = (x_{A1}, x_{A2}, x_{B1}, x_{B2})$ in which for at least one type (say A) the commodity bundles are not the same. That is

$$x_{A1} \neq x_{A2}.$$

Remember that this means that the *vectors* x_{A1} and x_{A2} are not equal in both their components.

There is clearly no loss of generality in labelling the individuals so that within each type the first is the, if anything, less favoured under the distribution x:

$$x_{A1} \precsim_A x_{A2}, \qquad x_{B1} \precsim_B x_{B2}.$$

Consider the bundle $\frac{1}{2}x_{A1} + \frac{1}{2}x_{A2}$ which is the simple average of what the two As are to get. By the convexity assumption on consumers' preferences **(3)**, it is A-preferred to x_{A1}:

$$x_{A1} \prec_A \tfrac{1}{2}(x_{A1} + x_{A2}). \tag{7.4}$$

We cannot take quite the same step in the case of the Bs, because we have not assumed that x_{B1} and x_{B2} are distinct bundles, as assumption **(3)** stipulates. However, if they are the same then $\frac{1}{2}x_{B1} + \frac{1}{2}x_{B2}$ is also the same, and so is indifferent to x_{B1}; if they differ, then we have a strict dispreference analogous to (7.4). So we can say that

$$x_{B1} \precsim_B \tfrac{1}{2}(x_{B1} + \tfrac{1}{2}x_{B2}). \tag{7.5}$$

Consider the coalition $\{A_1, B_1\}$ – the no-better-off A and the no-better-off B under the proposed distribution. If these two could redistribute their own endowments so that A_1 got the average A-bundle $\frac{1}{2}x_{A1} + \frac{1}{2}x_{A2}$ and B_1 got the average B-bundle $\frac{1}{2}x_{B1} + \frac{1}{2}x_{B2}$, then by (7.4) A_1 would do better than under the proposed distribution and by (7.5) B_1 would do at least as well as under it, so this coalition consisting of A_1 and B_1 would block, and the distribution x could therefore not be in the core.

Now x is feasible, so $x_{A1} + x_{A2} + x_{B1} + x_{B2} =$ the sum of all endowments $= 2e_A + 2e_B$. Therefore $(\frac{1}{2}x_{A1} + \frac{1}{2}x_{A2}) + (\frac{1}{2}x_{B1} + \frac{1}{2}x_{B2}) =$

$e_A + e_B$ = the sum of A_1's and B_1's endowments. So A_1 and B_1 *can* redistribute their endowments to get the average bundles. □

The Parity Theorem extends mechanically to any number of types, and to any number of each type.

Let us now consider trading games in which there are r players of each of two types – r As and r Bs – where r is any whole number $\geqslant 1$. Each A has preferences $\succsim_A$ and endowment e_A and each B preferences $\succsim_B$ and endowment e_B. We can immediately get two properties of the r-fold game.

(i) Consider any distribution x which is in the core of the r-game. In view of the Parity Theorem this distribution x is just some bilateral trade between a single A and a single B, replicated r times. We may think of this bilateral trade as a biatomic molecule and of x as a body consisting of r of these molecules. Perhaps A_1 makes this trade with B_{12}, A_2 makes it with B_5, and so forth.

(ii) Because the distribution x is in the core of the r game, it is not possible that A_1 and B_{12} could redistribute internally and do better than by the molecular trade; likewise with A_2 and B_5, and so on. Consider now an $r = 1$ game in which a single A has $\succsim_A$ and e_A and a single B has $\succsim_B$ and e_B; we shall call this the molecular game of which the original game is an r-fold replication. Then this single A and single B, being just like A_1 and B_{12} in every relevant respect, like A_1 and B_{12} can do no better for themselves than by the molecular trade of which there are r in x. So this molecular trade must be in the core of the $r = 1$ game.

We may therefore, in analysing the r-game, use the same two-person Edgeworth box diagram as before. Now it represents the trading possibilities of any dissimilar pair, say A_1, B_{12}. If the core or contract curve of this two-person molecule of the complete game is CC', we can be sure that a core distribution in the complete game is merely a replication of some point on CC'.

But can it be a replication of *any* point on CC'?

We saw in TU trading games that as the number of traders grew, the range of solutions, the range of elements in the core, got narrower. In those games the solution was defined in terms of payoffs – equal to the selling price for the sellers and to the difference between demand price and selling price for buyers. These variables were better and better determined as m and n grew, the intervals in which they had to lie closing down or *collapsing* on single points as m and n tended to infinity.

So it is here. As r increases, the core shrinks. As r tends to infinity, the core closes down on a single distribution. This distribution must, by our above arguments, be an r-fold replication of some point of CC' in the molecular, $r = 1$ game. This point is M.

Point M, recall, is the point of CC' at which the joint tangent of the A- and B-indifference curves passes through E. Because of the price-taking scenario for how M might come about, M may be regarded as the competitive equilibrium of the molecular game. This is the way in which we should understand the slogan 'as the number of traders increases without limit, the core shrinks to a competitive equilibrium'. Here, there *is no* given market price ratio. But as the number of traders grows, the traders are led inexorably by competition among their manoeuvring subgroups to behave *as if* there were. Not only that, but to behave as if there were a given market price ratio exactly equal to the slope of EM!

This is Edgeworth's famous limit theorem of 1881 (rigorously proved and generalised by Shubik, Debreu and Scarf, and others [24, 6].) In Edgeworth's words: 'Where the numbers on both sides are indefinitely large, and there are no combinations, and competition is in other respects perfect, contract is determinate.' Certain other conditions are in fact needed, notably convexity **(3)**. 'Combinations' means binding coalitions.

We now state and prove the

LIMIT THEOREM. Let G_1 be a trading game for two players A, B, with preferences $\succsim_A$ and $\succsim_B$ and endowments e_A and e_B, and let G_r be an r-fold replication of G_1. Then (a) the core of G_r is an r-fold replication of some subset of points of the core CC' of G_1. (b) As r increases the subset only sheds points, and never regains them. (c) As r tends to infinity, no point remains in it for ever apart from M, the competitive equilibrium of G_1. (d) M remains in it however large is r.

Point (a) we have already shown; (c) and (d) we have given notice of; and (b) is an additional result, which will be explained as we go along.

Instead of giving a formal and inevitably obscure proof, we demonstrate the theorem by a numerical example, which those who care to can easily generalise algebraically.

Figure 7.4 is the Edgeworth box for the molecular game G_1 and M is its competitive equilibrium. Consider any point D of CC' which $\neq M$. Call the A- and B-indifference curves which pass through

D, I_A and I_B, respectively. That some indifference curve of each passes through an arbitrary point follows from assumption **(2)** about consumers' preferences (continuity), which ensures that the whole space is as it were filled with indifference curves.

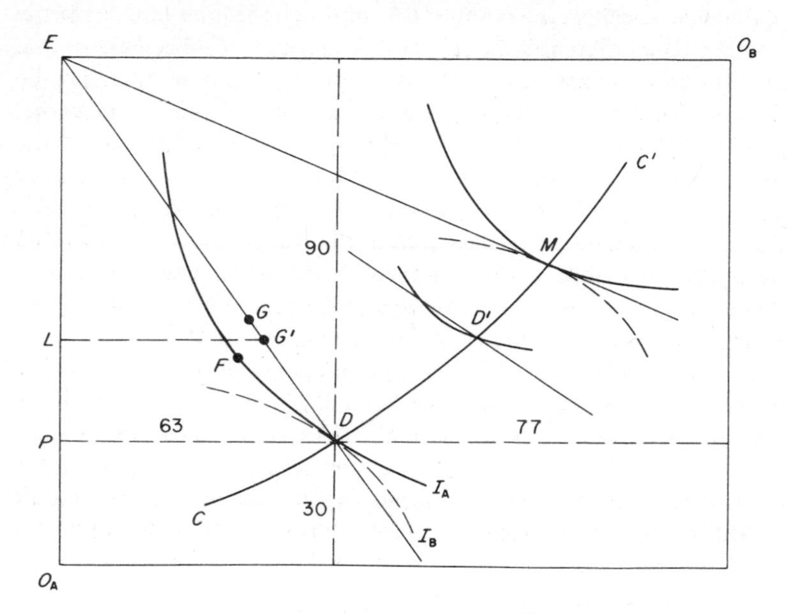

Fig. 7.4 *Edgeworth's limit theorem*

Now on ED, between E and D, there must be either points to the north-east of I_A or to the south-west of I_B. This is a consequence of the strict convexity of the indifference curves, assumption **(3)**. Say, without loss of generality, that there are points to the north-east of I_A, as in the case illustrated. Let G be any such point.

Now consider the ratio $EG : GD$, which is a number between 0 and 1, a fraction. Not all fractions are rational numbers, that is, ratios of whole numbers; for example, $\frac{1}{2}\sqrt{2}$ is not. But it is a fact that one can approximate any fraction – indeed, any number – as closely as one pleases by some rational number. So we can find a rational fraction which splits ED at a point close enough to G also to be on the north-east side of I_A. Say that G' is such a point and that it divides ED in the ratio 13 : 5. That is, G' is 13/18 of the way from E to D.

Concretely, let D be the trade in which A gives 90 of good 1 and receives 63 of good 2. Thus the distribution it determines is

$$x = \left[\begin{pmatrix}30\\63\end{pmatrix}, \quad \begin{pmatrix}90\\77\end{pmatrix}\right].$$

Consider now a many-trader game G_r. The point D, r times replicated, gives a feasible distribution in the game G_r. Denote this r-replication of x by $x^{(r)}$. In this distribution every A would end up with (30, 63) and every B would end up with (90, 77). The question is: is this distribution $x^{(r)}$ in the core of G_r?

Now, because G' is 13/18 of the way from E to D, 18 As trading from E to G' provide, and take, the same amounts of the two goods as 13 As trading from E to D – that is, the same amounts as 13 of the As trading in the proposed distribution $x^{(r)}$. But 13 As trading thus could put 13 Bs at D, since D is a feasible distribution. Therefore 18 As trading to G' can put 13 Bs at D.

Contemplate a coalition S consisting of 18 of the identical type-As and 13 of the identical type-Bs. If r is big enough – *viz.*, at least 18 – such a coalition can be formed in the large game G_r. Assume for the moment that it is formed. Then the 31 members of this coalition can make the exchanges we have just specified. But if they should, the 13 Bs in S would end up just where they would have been under the proposed distribution $x^{(r)}$ – at D; while the 18 As in S would be at G′ instead of at D. This the 18 As would prefer. Spelling it out, this is because G' contains more of both goods than points like F on I_A, on whose north-east side it lies; G' is therefore A-preferred to F by assumption **(1)** on consumers' preferences (insatiability), and therefore to D by transitivity. Thus, by internal trading of their endowments the members of S can unambiguously improve on their receipts under the proposed distribution $x^{(r)}$.

S therefore *blocks* $x^{(r)}$; D is not in the core of the r-game G_r.

[Check numerically the crucial step in the argument at which we claim that it is feasible for S to redistribute its endowments to put 18 As at G' and 13 Bs at D. Since G' is 13/18 of the way to D, EL is 13/18 of EP, and LG' is 13/18 of PD. So each A in going to G' gives up $\frac{13}{18}\times 90$ of good 1, and receives $\frac{13}{18}\times 63$ of good 2. Calculate the total amounts of 1 and 2 that are available for the Bs, and confirm that thirteenth parts of these amounts are 90 and 77.]

We have now done all the work of the proof and the results drop like fruits.

First, for *any* point of CC' different from M, there is some blocking coalition composed of n_A As and n_B Bs. In our illustration the segment of I_A cut off by *ED* was wide, so a rough approximation to *G* was still on this segment. We obtained this approximation, *G'*, by dividing *ED* in the ratio 13:18. The 13 and 18 became our n_B and n_A. For a finer approximation one needs, unless one is lucky, larger numbers. So had the segment been narrow we would probably have needed to argue in terms of a coalition with large n_A and large n_B. But as *r* gets indefinitely large this gets to be always possible, for then whatever are n_A and n_B, *r* eventually exceeds both. Thus, any point of *CC'* other than *M* is ultimately vulnerable. No point other than the competitive equilibrium is in the core of a trading game with a sufficiently great number of trades. This is the central result, part (c) of the theorem.

Second, as *r* grows the molecular core of the *r*-game either stays the same or shrinks. It can never recover lost points. For the coalitions which blocked them when *r* was lower can still be assembled. This is part (b).

It follows from these two results that the molecular core shrinks *towards M* progressively as *r* increases, and hence that points *near* to *M* are not eliminated until there are many traders. This can be seen in another way. For a point like *D'* the segment of *ED'* cut off by the A-indifference curve through *D'* is short. We have already argued that shortness of the cut-off segment means that to get a suitable '*G*' the numbers n_A, n_B may have to be large ones. Now we note that a stronger statement holds. One end of the segment is *D'* itself, so a ratio n_A/n_B which gives a suitable '*G*' must be near unity: for this, n_A and n_B *must* both be large (consider the series $\frac{1}{2}, \frac{2}{3}, \frac{3}{4}, \frac{4}{5}, \ldots$).

Third, *M* itself is never blocked, however many traders there are. For *EM* has no points to the north-east of the A indifference curve through *M*, nor to the south-west of the B one. Hence, we certainly cannot show by arguments of the above type that *M* is blocked. This, it is true, does not prove that *M* survives; but it strongly suggests it. A simple, independent proof due to Shapley (see [6]) is described in the next section.

7.5 THE GARBAGE DISPOSAL GAME – ANOTHER EMPTY CORE

Everything we have done in this last chapter is in one way uncharacteristic of game theory. We claimed at the beginning of the book that game theory can take in its stride cases which are intractable to classical methods and which classical economics is obliged, in our opinion disingenuously, to relegate to the category of anomalies. The theory of games has certainly been shown to have no inhibitions about admitting uncertainties and interdependencies. True, uncertainty seems to drop out of the picture as we pass from non-cooperative to two-person cooperative games. But it never really does, for it is always present in the non-cooperative undertow of every cooperative game. Interdependencies have always been in the centre of the stage, and they are in the *n*-trader game. To some extent therefore the model we have just finished with embodies the phenomena in whose analysis game theory has an advantage. But only to some extent. There is no risk – there has not been for some time. And there is only one kind of interdependence: my payoff depends on others' consumption only by virtue of the givenness of total supply. But another kind of interdependence is absent – there are no externalities in consumption; my payoff does not depend directly on the amounts consumed by other individuals.

The *n*-person trading game is by no means the last word of game theory in economics. We have spent some time on it because it is an important and much-noticed result. But it owes much of its fame to the fact that in it game theory seems to confirm, from a different direction, the belief of traditional microeconomics in the supreme place of competitive equilibrium. This appearance is quite deceptive. It is true that the shrinking of the core is a valuable addendum to the Walrasian and Paretian array of properties of perfect competition – a General on the Board. But the result of Edgeworth and Shubik *et al.* depends, like those properties, on a host of facilitating assumptions, economically meretricious. Of these, some, such as the assumption of no externalities, do less than justice to the analytical potential of game theory. Of others this cannot be said: two examples are the static assumption that the parameters of the problem, the preference relations and endowments, do not change through the period analysed; and the assumption of the players' perfect knowledge of each other's endowments and payoffs. Game theory relies just as heavily as neoclassical microeconomics on these.

The last game we shall look at tries to make amends for the first of these sorts of meretricious assumption. The author would not wish to be thought a lapsed game theorist.

The last part (d) of the Limit Theorem was proved independently by Shapley and is known as Shapley's Theorem. It is the proposition that in the above model of exchange, the *competitive equilibrium is in the core*. One way in which it may be read is as a strengthening of the famous proposition that a competitive equilibrium is Pareto-optimal. Being Pareto-optimal means: cannot be blocked by the coalition of all the players. Being in the core means: cannot be blocked by a coalition of any size.

The game we are about to describe – the last – is a trading game in which there are external effects in consumption. There are also other features which we have not yet met, peculiarities from the point of view of neoclassical theory but quite commonplace in reality, and quite amenable to game-theoretic analysis. The consequence of the externalities together with the other 'anomalies' is that the Shapley result breaks down dramatically: if there is a competitive equilibrium, it is *not* in the core.

In order to understand the havoc wreaked by the departures from standard neoclassical enabling assumptions, one first has to see how Shapley's proof works. The assumptions are weaker than those of the last section: there are just any old traders A, . . ., N – traders are not grouped into types. To show that a competitive equilibrium must be in the core, we suppose that some coalition S blocks a competitive equilibrium and produces a *reductio ad absurdum*. The main lines of the proof are as follows. Let the competitive equilibrium be the distribution $x = (x_A, \ldots, x_N)$ with associated price vector p (actually, only the price ratios matter). Let the redistribution of members' endowments which motivates S to block x give x'_K to individual K in place of x_K (for each K in S). We know that some K_0 in S prefers x'_{K_0} to x_{K_0} and the other Ks in S are no worse off. Now x is by assumption a competitive equilibrium, so for each K, x_K maximises K's utility given the prices. Hence, if x'_{K0} is preferable to x_{K0} it must be *more expensive* than x_{K0} at these prices (here the convexity assumption is used). Similarly x'_K must be at least as expensive as x_K at prices p for every other K in S. Therefore the (vector) sum of the x'_K over the Ks of S must cost more at prices p than the sum of the x_K in S. Now the latter must have the same value at prices p as the sum of the e_K in S, because in competitive equilibrium each individual is on his budget line. So we have: at prices p, $\sum_S x_K$ has a higher value than $\sum_S e_K$. But $\sum_S x'_K$ is the *same basket* as $\sum_S e_K$, since the x'_K in S are a redistribution of the e_K in S, and so it must have the same value. This contradiction proves the theorem.

THE GARBAGE DISPOSAL GAME [23]

The garbage disposal game departs in three ways from the standard assumptions: (i) the objects traded are bads rather than goods; (ii) there is free disposal; and (iii) this disposal is effected by dumping on other players without their leave. Conditions (i), which merely changes the direction of the preference in assumption **(1)**, does not matter in itself; (ii) does, because it means that there are feasible outcomes for a coalition which are not internal redistributions. This property becomes dynamite when combined with (iii). This combination produces the externality, namely: my non-consumption of the commodity brings about, whether you like it or not, a diminution of your utility. Notice that it is not just that there is more of the bad left over which someone has to consume: this would not be an externality, but merely an interdependence due to the givenness of total supply. Conditions (ii) and (iii) together mean that my consumption decisions can have direct effects on other particular players. (The game also has transferable utility, which means that there is no place for *strict* convexity of preferences. But this is not responsible for the breakdown of Shapley's result.)

Three people A, B and C are endowed with one bag of rubbish each. Let A have x_A bags in a new distribution and let his utility be

$$u_A(x_A) = -x_A.$$

Similarly for B and C. Garbage bags can be exchanged, but not destroyed. So whatever happens in the distribution their sum is preserved. Since utilities are linear in them and units of utility and bags are of equal size, the game is TU.

The exchanges may be made by agreement, or without agreement, by dumping. Furthermore, if I dump my bag on you, you cannot simply dump it back: we may suppose that I scatter its contents over your herbaceous border.

On these assumptions we have for the characteristic function:

$$v(I_3) = -3,$$

$$v(S) = -\mathrm{N}(\sim S) \qquad (S \subset I_n),$$

where $\mathrm{N}(\sim S)$ is the number of people not in S. For example, $v(\mathrm{A}, \mathrm{B}) = -1$.

We want to show that if there is a competitive equilibrium it is not in the core. Observe that this is certainly true if the core is empty.

If $x = (x_A, x_B, x_C)$ is in the core, then

$$u_A + u_B \geqslant -1, \qquad u_B + u_C \geqslant -1, \qquad u_C + u_A \geqslant -1,$$

therefore

$$u_A + u_B + u_C \geqslant -3/2.$$

But by property 6.4 of $v(I_n)$ in Chapter 6, $u_A + u_B + u_C = v(I_3)$; and $v(I_3) = -3$, so we have a contradiction. This shows the core is empty.

We have now shown what we claimed, that in this game, *if* there is a competitive equilibrium, it is not in the core – in other words it is not stable under assaults by coalitions of arbitrary size. Suppose it were the case, however, that in all games of which this is true there is no competitive equilibrium anyhow: the result would be academic. But this is not the case. The present game, for one, does have a competitive equilibrium. It is only one in name, however; since it is not in the core it could not arise out of competitive behaviour: the equilibrium price would have to be imposed.

Let p be a price paid for being relieved of one bag. In the following table the first column specifies the three possible trades for a player, the second the utility he gets from each of those, the third the values of p which make it best for him, and the fourth the excess supply that would result if he, and consequently everyone, chose it. It is assumed here that only whole bags are traded and that no one can accept more than one other's bag.

trade	*utility*	*price condition*	*excess supply*
dump own	$-p$	$p<1$	3
keep own	-1	$p=1$	0
take one	$-2+p$	$p>1$	-3

If $p = 1$, each could maximise his utility by keeping his own, and there would be equilibrium.

Finally, using this competitive equilibrium, let us check through Shapley's steps and see where the argument leading to membership of the core breaks down. Consider $S = \{B, C\}$. They consider $(\frac{1}{2}, \frac{1}{2})$ for themselves instead of $(1, 1)$. It is preferred all right. It is more expensive all right. It is practicable, by dumping. But it is *not* just a redistribution of their endowments $(1, 1)$, so we are not led to Shapley's contradiction.

References

[1] M. Allais, 'Le comportement de l'homme rationnel devant le risque: critique des postulats et axiomes de l'école americaine', *Econometrica* (1953).

[2] K. J. Arrow, *Social Choice and Individual Values* (Wiley, 1951).

[3] K. J. Arrow, *Essays in the Theory of Risk-Bearing* (North-Holland–American Elsevier, 1970).

[4] K. J. Cohen and R. M. Cyert, *Theory of the Firm* (Prentice-Hall, 1965).

[5] O. A. Davis and A. B. Whinston, 'Externalities, Welfare, and the Theory of Games', *Journal of Political Economy* (1962).

[6] G. Debreu and H. Scarf, 'A Limit Theorem on the Core of an Economy', *International Economic Review* (1963).

[7] G. de Menil, *Bargaining: Monopoly Power versus Union Power* (M.I.T. Press, 1971).

[8] J. T. Dunlop, *Wage Determination under Trade Unions* (Augustus Kelly, 1950).

[9] F. Y. Edgeworth, *Mathematical Psychics* (Kegan Paul, 1881).

[10] W. Edwards, 'The Theory of Decision Making', in *Decision Making*, ed. W. Edwards and A. Tversky (Penguin, 1967).

[11] W. Fellner, *Probability and Profit* (Richard D. Irwin, 1965).

[12] L. A. Festinger, *A Theory of Cognitive Dissonance* (Row, Peterson, 1957).

[13] M. Friedman and L. J. Savage, 'The Utility Analysis of Choices Involving Risk', *Journal of Political Economy* (1948).

[14] D. Gale, *The Theory of Linear Economic Models* (McGraw-Hill, 1960).

[15] T. Kuhn, *The Structure of Scientific Revolutions* (University of Chicago Press, 1962).

[16] F. H. Knight, *Risk, Uncertainty and Profit* (Houghton Mifflin, 1921).

[17] A. Leijonhufvud, *On Keynesian Economics and the Economics of Keynes* (Oxford University Press, 1968).

[18] R. D. Luce and H. Raiffa, *Games and Decisions* (Wiley, 1957).
[19] J. F. Nash, 'The Bargaining Problem', *Econometrica* (1950).
[20] J. F. Nash, 'Two-Person Cooperative Games', *Econometrica* (1953).
[21] P. A. Samuelson, 'Probability, Utility and the Independence Axiom', *Econometrica* (1952).
[22] L. J. Savage, *The Foundations of Statistics* (Wiley, 1954).
[23] L. S. Shapley and M. Shubik, 'On the Core of an Economic System with Externalities', *American Economic Review* (1959).
[24] M. Shubik, 'Edgeworth Market Games', in *Contributions to the Theory of Games*, ed. R. D. Luce and A. W. Tucker (Princeton University Press, 1959).
[25] M. Shubik, *Strategy and Market Structure* (Wiley, 1959).
[26] H. A. Simon, 'A Behavioral Model of Rational Choice', *Quarterly Journal of Economics* (1955).
[27] E. von Böhm-Bawerk, *The Positive Theory of Capital*, trans. W. A. Smart (Stechert, 1930).
[28] J. von Neumann and O. Morgenstern, *The Theory of Games and Economic Behavior* (Princeton University Press, 1944).
[29] L. Wittgenstein, *Philosophical Investigations* (Blackwell, 1953).
[30] F. Zeuthen, *Problems of Monopoly and Economic Welfare* (Routledge & Kegan Paul, 1930).

Index

THE POLYTECHNIC OF WALES
LIBRARY
TREFOREST